FROZEN
FURY

AGRICULTURAL CROPS
and
HAIL DAMAGE

M.G. BUCHOLTZ
B.Sc., MBA, M.Sc.

A Wood Dragon Book

Frozen Fury:
Agricultural Crops and Hail Damage

Published by :
Wood Dragon Books,
Box 429, Mossbank, Saskatchewan, Canada, S0H 3G0
http://www.wooddragonbooks.com

ISBN:
978-1-989078-85-3 paperback
978-1-989078-92-1 e-Book

Contact the author: supercyclereport@gmail.com

TABLE OF CONTENTS

INTRODUCTION

In late 2017, I made the decision to go to graduate school at Heriot Watt University in Edinburgh, Scotland. My goal was to deepen my knowledge of brewing and distilling science so I could provide more in-depth consulting services to start-up craft distilleries.

My thesis project centered around distilling vodka from an ancient-grain wheat varietal called *CDC Prime Purple Wheat*. In early 2020, as I began writing my thesis paper, I realized that my classroom learning had not provided me with a complete understanding of cereal crop plant growth. And that is when things took an unexpected turn for me.

In March 2020, I spied an ad in the Assiniboia Times newspaper from a hail insurance company looking for crop hail insurance adjustors. I reasoned that if I worked as a hail adjustor during the summer of 2020, my practical knowledge of cereal grain plant growth would be greatly expanded. I could then incorporate this knowledge into my thesis paper. I emailed my resume to the insurance company; two weeks later I was interviewed over the phone (Covid restrictions did not allow for a face-to-face meeting), and offered a position as a hail adjustor trainee.

In July 2020, I attended a 2-day training session with a dozen other new-hires. We were introduced to the effects of hail damage on cereal, pulse and oilseed crops. A couple days before this training was slated to start, the Assiniboia area in southern Saskatchewan was blasted with a severe hail storm. Shortly after the training session, I was assigned to a veteran adjustor for in-the-field training. For 12 days, I followed in his steps as we examined durum wheat, oats, barley, lentils, canola and even some corn. I became fascinated with the fury of nature and how frozen ice pellets could render so much damage to plants. After my training period, I was told to turn my focus to passing the Hail Adjustor Exam (I was told the exam would be multiple choice format with 75% needed to pass). In late 2020, I passed my exam with a mark of 86% and became a licensed hail adjustor. My sights were then set on the 2021 crop year.

As the 2021 crop year got underway, the weather projections for Saskatchewan were suggesting a hot dry summer with little storm activity. I began to wonder if I would even be called out for any adjusting activities. Then all hell broke loose. Hail storms arrived in earnest across many parts of Saskatchewan, Alberta, and Manitoba. I ended up completing 52 days of adjusting activity that took me to parts of Saskatchewan I have never been to before.

Walking several miles each day through fields gave me ample opportunity to ponder the meteorological origin of hail storms. Walking through farm fields looking at damaged crops also heightened my curiosity about seed structure, stages of crop growth, and plant recovery from hail damage.

In late September when I finally returned home I decided to seek answers to some of the questions that were burning in my mind. As I looked through my collection of books in my office, I found the textbook I had used 42 years ago in pilot training school. I was 16 years old at the time and had wild-eyed ambitions of becoming a top-gun fighter pilot in the military. I thought that obtaining my private pilot license would be a good first step for me. In the end, Engineering school at Queen's University ultimately beckoned louder than Royal Military College, and life took a different turn. As I sifted through this old flight training text, I found the comprehensive chapter on cloud formations and weather than can affect a small aircraft. And there it was – an explanation of how hailstorms develop.

My hunger for more knowledge quickly intensified. My M.Sc. degree studies at Heriot Watt University left me with long-term access to their online Library. When I entered queries such as 'pulse crop growth' and 'cereal crop growth', I was rewarded with hundreds of peer-reviewed journal studies from institutions around the world. When I entered search terms such as 'crop hail damage' I was overwhelmed with scientific information on hail storm dynamics. The literature made it clear that hail affects everything from onions in Spain to grapes in Argentina to watermelons in Texas. Any thoughts on my part that hailstorms were exclusively a North American phenomenon were quickly dispensed with. I further discovered that academic studies involving both simulated and actual hail damage have been shared with crop insurance companies around the world who have used the data to quantify their indemnity loss calculations.

The overwhelming amount of hail-related literature left me with a nagging sense that there was a void to be filled. My editor at Wood Dragon Books, who has published my many other books, posed the questions: what if I could create a book that would offer a detailed technical explanation of how hailstorms develop? What if the book could describe the various stages of plant growth for a wide variety of crops? What if it could describe academic research studies on the effects of hail damage on various crops? And most importantly, could the book could be written with farm operators, hail adjustors, classroom students, and just about anybody interested in agriculture in mind?

These what-ifs have now taken on a solid reality. Each of the chapters in this book contains deep scientific knowledge. But each chapter also ends with a summary of the chapter material for the benefit of the busy reader on the go. Whatever your connection to agriculture and whatever your available reading time, I hope this book refreshes and expands your knowledge of crop growth, hail storms, and crop damage. If you have ever thought about becoming a hail insurance adjustor, I further hope this book will inspire you to follow through on those thoughts.

CHAPTER 1
WEATHER, CLIMATE, AND CHANGE

Weather

Understanding the science of hail hinges on understanding the basic science behind our weather. But what is *weather?*

Weather is defined as the state of the atmosphere in a particular area at a given moment in time. Multiple times every day, whether on radio, television or the internet, we hear or read about the weather. Temperature, wind direction, wind speed, and barometric pressure all help to define the state of the atmospheric conditions in our area at that time.

A *weather forecast* is a best guess of what the atmospheric conditions will be one, two, three, or more days in the future.

Climate

Climate is defined as the weather conditions over long periods of time. If you think about the weather in your province or state over the past years and decades, it should be apparent that there has been a change. The weather today is different than the weather 30 years ago. This is where the expression *climate change* comes from. Atmospheric conditions (the weather) are becoming more volatile and more unpredictable. Droughts are more severe and more widespread when they occur. Storms are getting more ferocious. Storms are delivering hail more often. Hail damage to crops is increasing in severity. A 2018 paper by researchers at the National Centre for Atmospheric Research in Colorado, USA shows large-hail-day events have tripled in frequency since 1979 in the US, Australia, and central Europe. [1]

Climate Change - cyclical or man-made?

In 2019, as I was finishing off my M.Sc. degree studies at Heriot Watt University, I took an elective course in Renewable Energy. My goal in taking this course was to learn more about wind turbines and solar panels and their roles in powering the future economy.

My course experience revealed something unexpected. The young professors who taught the various modules of the course sincerely believed that the entire world is due to end in 2030 unless world governments fully embrace green energy and make a hard left turn away from fossil fuels. Throughout this course not a word was uttered about atmospheric emissions from natural disasters such as forest fires or volcanic eruptions. The focus was exclusively on emissions from mankind burning fossil fuels. I learned that this is called *anthropogenic* climate change. I found myself reluctant to accept their opinions at face value.

With my studies done and my thesis written, I decided to investigate the climate change subject deeper. I found the textbook from the Geology 101 class I had taken at Queen's University in 1982. Were we taught about climate change back then? And there it was! A 1976 academic study by authors Hays, Imbrie and Shackleton published in the journal *Science* under the title *Variations in the Earth's Orbit: Pacemaker of the Ice Ages.* Academics, such as the ones at Heriot Watt University who taught the Renewable Energy course, with a fixation on anthropogenic climate change, have seemingly chosen to ignore this critical study. (2)(3)

Hays, Imbrie and Shackleton were granted access to drill core material from two locations in the southern Indian Ocean. This core material had been obtained as part of a multi-nation study called *Climate: Long Range Investigation, Mapping, and Prediction* (CLIMAP), which had been conducted in the early 1970s.

In their analysis of the drill core material, Hays, Imbrie and Shackleton focused on the Milankovitch Cycles. These cycles were proposed through the research of Serbian astrophysicist Milutin Milankovitch. Born in 1879, Milankovitch graduated from Vienna Technical University in 1904 with a doctoral degree. He spent his entire career researching and teaching at Belgrade University in Yugoslavia (modern day Serbia). He determined that Earth is tilted on its axis by about 23.5 degrees. This angle of tilt varies from 22.1 to 24.5 degrees over a cycle of about 40,000 years. He noted Earth orbits the Sun in a pattern that is more elliptical than circular. Variations in this eccentricity unfold over a long cycle of around

7

100,000 years. Lastly, he noted Earth rotates (spins) on its axis with a slight wobble. The cycle of this wobble measures 25,772 years in duration. [4]

What Hays, Imbrie and Shackleton concluded from examining the drill core data was that climate change is cyclical. The data pointed to three long cycles that overlap with one another in a complex fashion. The data showed these long cycles were very closely aligned to the cycles identified by Milankovitch.

To further appreciate climate change, stop for a moment and consider that there are oil and gas deposits in the North Sea off the coast of Norway. There are massive untapped oil and gas deposits in northern British Columbia that extend into the nearby North West Territories. In order for these deposits to have developed, millions of years ago these areas must have been warm and wet with plenty of decaying vegetation that transformed into hydrocarbons. Mankind was not around millions of years ago to cause these warm, wet conditions. Therefore, the findings of Hays, Imbrie and Shackleton are significant. Where we currently are within the larger Milankovitch cycles confirmed by Hays, Imbrie and Shackleton remains uncertain.

The population of Earth in the early 1900s when Milankovitch was studying cycles was around 1.6 billion people. Today, the population of Earth is approaching 7.9 billion people. Therefore, climate change might be a combination of Milankovitch cycles and anthropogenic factors. The extent to which cyclicality and anthropogenic factors each are contributing to climate change cannot be quantified with accuracy. Scientists do not have enough long-term data to make these distinctions. Consider that the first reasonably accurate thermometer was not invented until 1715. It was not until around 1880 that mankind started to accumulate temperature data on a regular basis. The amount of available temperature data is less than 150 years-worth. To put this in context, the last Ice Age peaked some 20,000 years ago. Our planet is over 4 billion years old.

Society at large remains divided as to the cause of climate change. A 2021 survey conducted by Yale University [5] revealed: respondents in Spain (64%) and Italy (60%) were the most likely to think that climate change is mostly caused by human activities. Respondents in Indonesia (16%) and Nigeria (24%) were the least likely to think that climate change is mostly caused by human activities. In the US, 40% of respondents thought climate change is mostly caused by human activities. Canada recorded a higher response at 50%.

While opinion differences will likely persist, one thing is for certain. Society will have to adapt to climate change.

Hot Air Rises, Cool Air Sinks

Our planet is a sphere. We owe our survival to the warmth of the Sun's rays hitting the planet. But as Figure 1-1 shows, the Sun's rays do not strike Earth in a uniform fashion.

Because our planet is a sphere, the Sun's rays will hit the area near the equator at a 90-degree right angle. These rays impact a small geographic area and create warmer weather conditions.

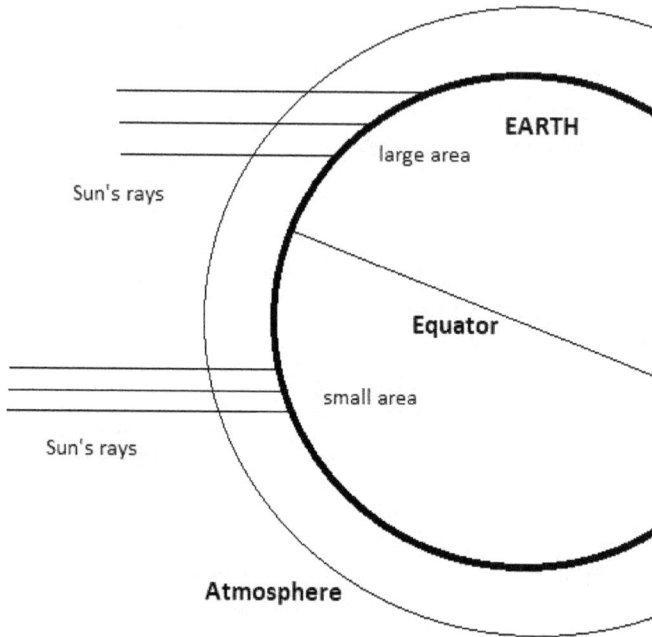

Figure 1-1
Sun's rays striking Earth

The rays that hit the parts of Earth above and below the equator cover a larger area and hit at an oblique (less than 90-degrees) angle. Because these rays have travelled a greater distance through the atmosphere, they have lost heat energy along the way. As a result, they impart less heat energy to the planet. This rudimentary explanation accounts for the hotter temperatures near the equator and the cooler temperatures at higher latitudes. This is why we seek out winter vacation opportunities in southerly locations as opposed to northerly locations. This is why we have cooler summer-time temperatures in Alberta than in Mexico.

Figure 1-1 further denotes a feature called the *atmosphere*. The atmosphere is comprised of several individual layers: the *troposphere (closest to Earth)*, the *stratosphere*, the *mesosphere*, the *thermosphere*, the *ionosphere*, and finally *space*.

The weather that we experience is created in the troposphere. This component layer of the atmosphere ranges in height from about 8500 meters (28,000 feet) at the North and South poles to about 16,000 meters (54,000 feet) at the equator.

As an aside, the world witnessed two flights into *space* in September 2021 (Space-X and Blue Origin). The technical definition of space is the *von Karman Line*. This is the point calculated as being 100 kms (62 miles) above Earth, where the air is so thin that a winged aircraft cannot remain aloft. Despite the conflicting opinions in the media, each of these celebrity launches did in fact reach space.

Planetary Rotation and Cells

We intuitively know that hot air rises and cool air sinks. If Earth was stationary, the heated air from the region near the equator would rise and flow to higher latitudes. It would eventually cool and sink back towards the equator. It would then heat again thanks to heat energy from the Sun's rays and the cycle would repeat.

But Earth is not stationary. Nor is it of uniform construction. The distribution of land and sea varies at different points around the planet. Furthermore, Earth rotates in a west-to-east direction. To a theoretical observer hovering above the North Pole, Earth appears to be rotating counter-clockwise. This rotation combined with the non-uniform land and sea distribution adds several layers of complexity to the flow of heat energy from the Sun's rays. This complexity leads to a discussion of *Cells*.

Hadley Cells and Polar Cells

Warm air accumulating at the equator rises to a height of up to about 18 kilometers (11 miles) above Earth. As the warm air rises, it spreads northerly and southerly on either side of the equator.

By the time the warm air has reached about 30 degrees of latitude on either side of the equator, large portions of the warm air have cooled. The cooled air starts to sink, and flows back to the equator region. This circulation pattern is called a *Hadley Cell*, named after George Hadley (1685-1768). Hadley was an English lawyer, educated at Oxford. He was also an amateur meteorologist who was intrigued by the wind forces that helped ships sail from Europe to the Americas.

Not all of this warm air sinks and flows back to the equator. Some of it remains aloft and accumulates at the region 60 to 70 degrees of latitude at about 8 kilometers (5 miles) above the surface of Earth. In each hemisphere, this accumulated warm air eventually heads towards the poles. As the air heads poleward, it cools and begins to sink and head back towards 60-70 degrees latitude. This thermal-polar circulation pattern is called a *Polar Cell*.

Rossby Waves and The Jet Stream

The motion of air in the troposphere is normally at the same general latitude. But this flow can often shift into a pattern of multiple waves, called harmonic waves, or *Rossby Waves* (named after Swedish-born meteorologist Carl-Gustav Rossby).

Rossby waves, along with the rotation of Earth, create movements of air in patterns called the *jet streams*. Weather forecasters tend to lump jet stream flows together and refer to a singular 'jet stream'. In fact, there are four jet streams – two in the northern hemisphere and two in the southern hemisphere. In each hemisphere there is a jet stream at about 30 degrees of latitude (sub-tropical jet stream) and another at 60 to 70 degrees of latitude (polar jet stream). A jet stream can best be visualized as a flowing river of air. This river of air can be hundreds of kilometers wide and a few kilometers deep.

The jet stream was first encountered on a practical level in 1934. American aviator, Wiley Post, was noted for his high-altitude flying experiments. On December 7, 1934, he took his aircraft to 20,000 feet and encountered a wind force that greatly accelerated his air speed. This was the jet stream.[6]

Pressure Gradient, Wind, and Coriolis Force

To further understand the jet stream, envision 2 columns of air, one cold and the other warm. The warm air is from lower latitudes and the cold air from higher latitudes. The column of cold air is denser and exerts less pressure. Intuitively, we have experienced this if we have had a tire on our vehicle appear to be underinflated when the temperature gets cold in winter. We also know intuitively that warm air will flow towards cool air as Nature seeks a stable balance. Warm air flowing from the warm column towards the cold column creates a *Pressure Gradient*. The movement of air within a pressure gradient is called *wind*. As the warm air flows northward towards the cold air, the flow is deflected off to the right by the counter-clockwise rotation of Earth. The opposite deflection occurs in the southern hemisphere. This deflective force is termed the *Coriolis Force*.

A Shifting, Looped Ribbon

The variations in land and sea masses around the globe affect the Rossby waves and the pressure gradient. The result is jet streams that take on the appearance of ribbons with loops and bends as shown in Figure 1-2.

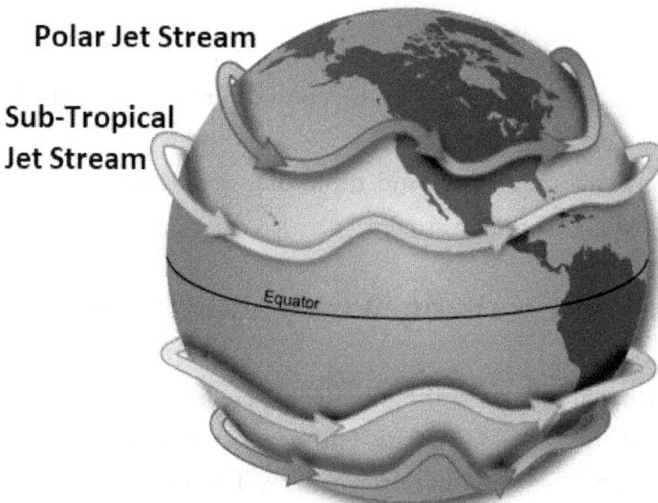

Figure 1-2
Jet stream(s)

Although the jet stream(s) will shift northerly or southerly in winter and summer respectively, it always flows west to east at speeds up to about 450 km/hr (275 miles per hour) and at a height of 4 to 8 miles (7-13 kilometers) above ground. If you have taken a direct flight from a place like Vancouver, easterly to a European destination, you will have likely experienced the jet stream pushing the airplane and creating an earlier than expected arrival in Europe. Your westerly flight coming home to North America will have taken longer because the aircraft was fighting the flow of jet stream air.

Strong and Weak Jet Streams

A *strong jet stream* is comprised of winds that are intense and compressed. A strong jet stream has fewer loops and bends and is situated closer to the northern latitudes. Warmer air from southerly latitudes will then flow northward as Nature seeks a balance. This flow can create hotter and stormier weather conditions. In places like India, a strong jet stream can lead to a more intense monsoon rain season.

A *weak jet stream* is comprised of weak winds. A weak jet stream has an increased number of loops and bends. A weak, irregular, jet stream can suddenly dip southwards and bring cold air and snow to places like Texas. In February 2021, all counties in Texas were blasted with snow. The last time such widespread weather hit Texas was 126 years ago. A weak jet stream can also dip southerly over the Gulf of Mexico, pick up moisture, turn northerly again, and dump that moisture as rain or snow over the US Great Plains states.[5] Stronger and weaker jet streams are all part and parcel of climate change. Going forward, it seems reasonable to expect more erratic jet stream behaviour.

Low Pressure and High Pressure

The top parts of the jet stream ribbon are called *ridges*. The lower parts of the jet stream ribbon are called *troughs*. The west-to-east air flow will accelerate around the ridges and troughs in the jet stream pattern.

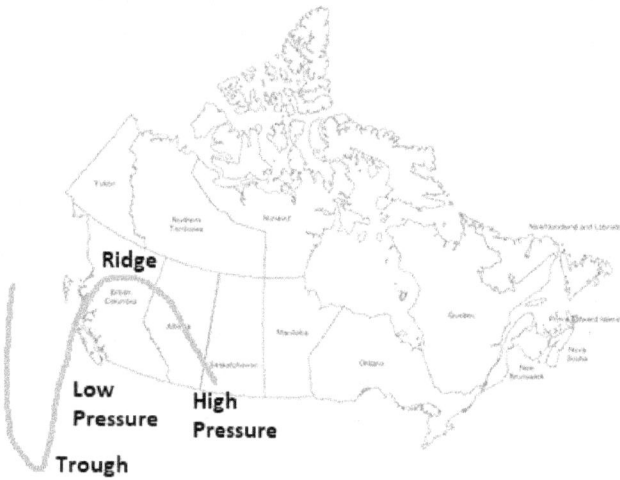

Figure 1-3
Ridges and Troughs

As Figure 1-3 illustrates, an area of *low pressure* will develop at a point east of a trough and mid-way to a ridge. Winds in a low-pressure area will move counter-clockwise. Warm, moist, southerly air moving to fill the low-pressure area will converge and rise, leading to cloudy, windy, showery weather. An area of *high pressure* will develop east of a ridge. Winds in a high-pressure area will move clockwise. As an example, consider that in November 2021 a trough area developed off the British Columbia coast. The resulting low pressure area that developed was situated over southern parts of the province which were hit with devastating rainfall and flooding. The associated high pressure area brought clear skies and above normal temperatures to Alberta and Saskatchewan.

Ferrell Cells –Linking Hadley and Polar Cells

To complete our examination of Cells, there is one more feature to consider. Situated between the Hadley Cells and Polar Cells are *Ferrell Cells* named after American meteorologist William Ferrell (1817-1891). Ferrell Cell circulation occurs between about 30 degrees and 60 degrees of latitude. Part of the rising Ferrell Cell air at 60 degrees latitude feeds the formation of the Polar Cells.

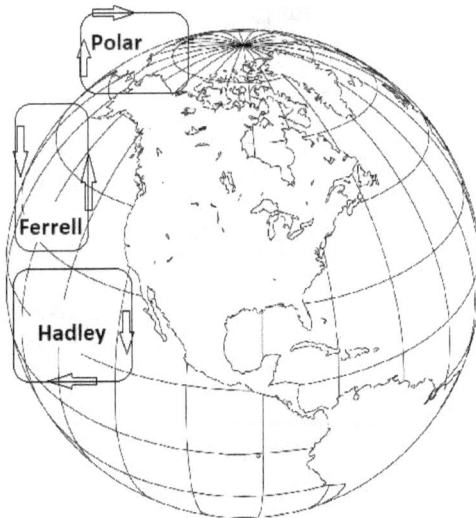

Figure 1-4
Polar, Ferrell, and Hadley Cells

The remainder of the Ferrell Cell air descends towards the equator where it collides with the Hadley Cell air at high altitude. Ferrell Cells are not air temperature related. They derive from the circulatory action of the Polar Cells and the Hadley Cells. Ferrell Cells circulate in an opposite direction to the Hadley and Polar Cells, imparting balance to the atmosphere's circulation. The sketch in Figure 1-4 illustrates the Polar, Ferrell, and Hadley Cells.

Also, the action of the Ferrell Cells creates semi-permanent areas of high and low pressure around the globe. In semi-permanent areas of low pressure, more rainfall will be encountered. In semi-permanent areas of high pressure, less rainfall will be encountered. This explains, for example, why the climate in Arizona, USA is drier than the climate in Alberta, Canada. This explains why parts of South America and Africa have vast regions of rain forest. Figure 1-5 illustrates the semi-permanent areas of the globe that arise due to the Ferrell Cells. (6)

Figure 1-5
Low Pressure Areas related to Ferrell Cells

Hadley and Ferrell Cells act to create wind patterns known as the *Trade Winds*. Air flow from a Hadley Cell returning towards the equator is shifted in a westerly direction due to the counter-clockwise rotation of the Earth. This creates a wind flow from east to west in the equatorial region. The wind flow above the equator is called the *North East Trade Winds*. Its southerly component is the *South East Trade Winds*. These Trade Winds filled the sails of ships being commanded by early explorers such as Columbus and pushed them to tropical destinations from their starting point in Europe. Figure 1-6 illustrates these flows.

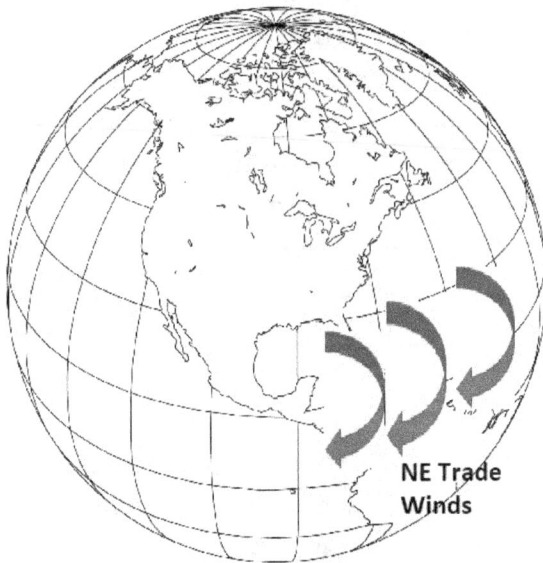

Figure 1-6
NE Trade Winds

El Nino and La Nina

Trade Winds do not blow regularly nor do they blow at constant speed. There are time intervals, that can be up to several years in duration, when the movement of the Trade Winds is interrupted. Scientists remain uncertain as to why these intervals occur in such an irregular fashion.

One of these interruption periods is called *El Nino*. When the flow rate of the Trade Winds slows, the sea surface temperatures in the eastern Pacific Ocean warm slightly. The fish, who prefer cooler waters, will stay at depth and fishermen off the coast of Peru and Chile notice a reduction in the size of their catches. But the effects of El Nino are more far-reaching than just fishing. Weather patterns can turn wetter along the west coast of North America and drier in south-east Asia, India and Australia. Climatologists will declare an El Nino event to be in effect when surface water temperatures in south-east Asia rise to more than 0.5°C (1°F) above the long-term baseline temperature for three consecutive months. El Nino events have recently been recorded in: 2002-7, 2009-10, 2014-16, and 2018-19.

The counterpart to El Nino is *La Nina*. A strengthening of Trade Winds causes warmer waters off the west coast of South America to migrate towards south-east Asia. The net effect is an increase of convective activity with cooler, deeper water coming to the surface, bringing with it more fish. Fishermen in places like Peru reap the benefits. Fishing aside, La Nina events lead to increased rainfall in Australia, India, and south-east Asia. Large parts of the USA will experience drier weather patterns. La Nina events have recently been noted in: 2000-1, 2008-9, 2010-11, 2016-17, and 2020-21.

Scientists are in agreement that during 2020 and into early summer 2021 La Nina dry conditions existed across North America. The dismal crop progress numbers reported by the USDA in the US northern Great Plains along with the poor crop conditions that I personally observed during the Summer of 2021 support the existence of a La Nina event. As of late October 2021, scientists are suggesting an 87% probability of a La Nina event developing in the latter months of 2021 and persisting through 2022. Should this La Nina situation come to pass in 2022, the implication is continued drought and possibly more hailstorm events.

Arctic Amplification

The increasing global population is also affecting the physics of weather. As the United Nations reminds us, the global population stood at 5 billion in 1987 and 6 billion by 1999. In 2011, the global population had reached 7 billion. The world's population is expected to increase by 2 billion persons in the next 30 years to reach 9.7 billion by 2050.[7] More people mean more emissions. The Center for Climate and Energy Solutions (C2ES) estimates that in 2021 the world collectively emitted just under 35 billion tonnes of carbon dioxide. In 1990, this figure was just over 20 billion tonnes.[8]

Growing population and increasing carbon dioxide emissions (anthropogenic factors) are contributing to warming in the Arctic region and melting of the polar ice. No doubt the polar ice is also melting due to variations in Earth's tilt, orbital path, and wobble as outlined in 1976 by Hays, Imbrie and Shackleton. Whatever the combination of causes, a decrease in solids (ice) and an increase in liquids (seawater) in the Arctic regions is occurring.

The research of Dr. Lei Wang of Harvard University is focused on the warming waters in the Arctic region. He refers to this phenomenon as *Arctic Amplification*. Wang argues that as Arctic ice quantities are reduced, the Arctic waters take on more solar heat energy. This energy serves to thicken the troposphere over the northern latitudes. This, in turn, affects circulations in the Hadley, Polar, and Ferrell cells. The affected circulations influence the Rossby waves in the jet streams. A thicker troposphere at higher latitudes stands to reduce the North-South pressure gradient. This, in turn, stands to weaken the jet streams and promote more erratic weather in mid-latitude areas. The implications for agriculture in western Canada and the US Great Plains states are profound. More erratic weather is not conducive to good crop yields. More erratic weather can lead to a greater frequency of hail events. Wang exemplifies his weaker jet stream scenario by pointing to the abnormal winter events seen in Texas in early 2021. In my opinion, the weather conditions and hail events that I witnessed in 2021 also exemplify Wang's argument. [9][10][11]

The weather is changing, the climate is changing, the world is changing. We can no longer assume weather will behave in a predictable manner. Hadley, Polar, and Ferrell Cells are being disrupted. Trade Wind flows are also being disrupted. La Nina and El Nino events are increasing in frequency. The jet streams are behaving more erratically. Looking forward, when it comes to weather, expect the unexpected.

To Sum It Up

» *Weather* is the state of the atmosphere at a particular location at a given moment in time.

» Weather conditions over long periods of time are defined as *climate*.

» Climate is changing. This is due to a combination of: variability in the Earth's orbital pattern, natural disasters such as forest fires and volcanic eruptions, and *anthropogenic* (man-made) emissions.

» Warm air from the equator region rises to about 30 degrees of latitude in a *Hadley Cell* pattern.

» Some of this warmed air rises to about 60 degrees of latitude in a *Polar Cell* pattern.

» Air flow in the Polar Cells contributes to the *Jet Stream*.

» The Jet Stream is like a ribbon of flowing air with loops and bends.

» The top parts of the looped jet stream are called *ridges*.

» The lower parts of the looped jet stream are called *troughs*.

» Strong Jet Stream flows can lead to hotter and stormier weather.

» Weaker Jet Stream flows can lead to erratic weather across North America.

» *Ferrell Cell* flow provides a balance between Hadley and Polar Cells.

» Hadley and Ferrell Cells in conjunction with the Earth's rotation create the *Trade Winds*.

» Variability in Trade Wind flows leads to *La Nina* and *El Nino* weather patterns.

» Increasing global population along with climate change is causing a reduction in Arctic ice quantities.

» As the Arctic region warms, Hadley, Polar, and Ferrell cells will be further affected. The Jet Stream(s) stands to weaken and become more erratic. Weather will become more extreme. The impact on global agriculture will be significant.

CHAPTER 2
HAIL

The Physics of Weather Fronts

When a large air mass sits over a geographic area for a couple days the air mass will pick up the temperature and humidity characteristics of the geographic area.

High altitude winds can push these air masses to neighbouring geographic areas. If an air mass with certain temperature and humidity characteristics is pushed over a geographic region with different characteristics, the result will be a *weather disturbance*.

Cold air masses are heavier than warm air masses. Therefore, high altitude winds will cause a cold air mass to move slower than a warm air mass. Conversely, a warm air mass will move faster than a cold air mass. If a faster-moving warm air mass begins to interact with a slower-moving cold air mass, the warm air mass will rise over top of the cold mass. This type of disturbance is referred to as a *warm front*. Any moisture contained in the warm air mass will precipitate out as the air rises. The precipitated moisture will appear in the sky as high-level clouds.

A moving cold air mass bumping against a stationary warm air mass will cause the warm air mass to rise. Moisture in the warm air mass will precipitate and form larger, more billowing clouds. This type of disturbance is referred to as a *cold front*.

Cloud Formations

Meteorologists classify clouds by their shape, content and mode of formation. Meteorologists generally divide clouds into four formations: *high clouds*, *medium (middle) clouds*, *low clouds*, and *convection clouds* (vertical development). The diagram in Figure 2-1 illustrates these various cloud formations.

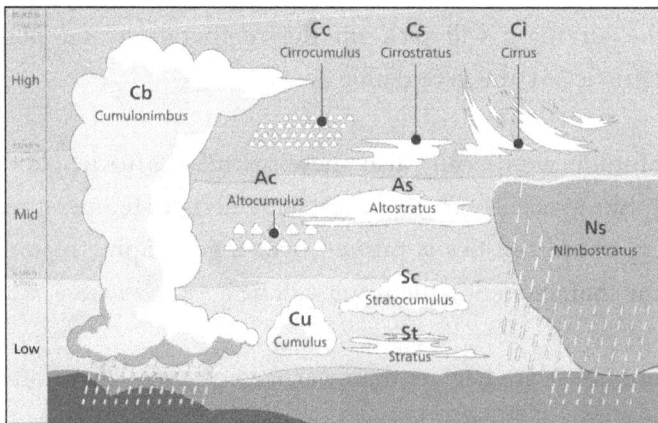

Figure 2-1
Cloud Formations

High Clouds

Cirrocumulus, Cirrus and Cirrostratus clouds comprise the high cloud category. These formations are described as 'wispy', 'feather-like', or 'cotton-like'. These clouds are associated with warm fronts and are early warnings of the potential for deteriorating weather.

Medium (Middle) Clouds

Altocumulus and *Altostratus* clouds comprise the medium (middle) cloud category. Altostratus clouds are also associated with warm fronts.

Low Clouds

Stratus, Stratocumulus, and *Nimbostratus* clouds comprise the low cloud category. These puffy clouds are usually between 600 to 900 meters (2000 to 3000 feet) above the ground. These formations are associated with the development of cold fronts.

Convection Clouds (Vertical Development)

Cumulus and *Cumulonimbus* clouds comprise this cloud category. A cold front formation is associated with this cloud category. Cumulus and Cumulonimbus clouds develop by accumulating moisture. Up-drafting air currents, called *thermals*, complete the process. Cumulus clouds are generally described as 'cotton balls'. Cumulonimbus clouds are towering, massive structures often with an anvil-like top. These are called *storm clouds* or *storm cells* and are often the origin of hailstorm events. [1]

The Physics of Hail

Hailstorms are a unique scientific phenomenon in that no scientist has ever been able to sit inside a storm cloud to observe the process of hail formation. Hailstorms are further unique in that owing to the dynamics of weather, no two hailstorms are alike. As a result, scientists have developed explanations and models for hail formation based on rudimentary thermodynamics, chemistry fundamentals, weather balloon data, radar data, and atmospheric data from airplanes flying passes through storm clouds. These explanations and models started to gather momentum in the early decades of the 1900s. These explanations and models continue to be discussed, debated, revised, and expanded to this day. More sophisticated and evolving technology has lent added momentum to the effort to better understand hail events. The physics of hail is a dynamic subject area. Despite the many facets of hail storms that are routinely debated and studied, scientists agree that there are three key ingredients that a storm cloud must provide in order to produce hail. These ingredients are: a nucleation site (called an *embryo*), supercooled moisture, and growth time. The WBF Model and the work of F.H. Ludlum focus on these key ingredients.

Wegener-Bergeron-Findreisen (WBF) Model

The period 1910 through the late 1930s was a vibrant time for research into cloud formations. German scientist Alfred Wegener, Norwegian scientist Tor Bergeron and Czech scientist Walter Findreisen are collectively credited for their early 1930s model describing nucleation of ice crystals in cloud formations. The inspiration for their collaborative efforts started with an observation of hoar frost forming on tree branches. They reasoned that tiny nucleation sites on the tree branches attracted moisture from the air with the net result being the formation of tiny, frozen ice crystals.

The WBF Model hinges on the 1st Law of Thermodynamics which states $\Delta U = Q-W$. In plain English, the internal energy of a system, (ΔU) equals the heat energy of the system (Q) minus the work done (W) by the system on the surrounding atmosphere. The WBF Model asserts that a cloud formation is a thermodynamic system. Moreover, the WBF Model asserts that a storm cloud is an *adiabatic system*, meaning no heat energy is added to the cloud system. That is, heat energy (Q) in the equation $\Delta U = Q-W$ equals zero. For an adiabatic cloud formation, $\Delta U = -W$. The minus sign in front of the work (W) factor means that the cloud formation does work on the surrounding atmosphere by making the cloud grow bigger.

The thermodynamic aspect of the WBF Model is the result of Findreisen's research that focused on the growth of liquid droplets and ice crystals in a cloud formation. Building on the concept that a cloud formation as an adiabatic system ($\Delta U = -W$), he further reasoned that as up-drafting air (thermals) rises to merge with the cloud formation, the rising air adiabatically cools. The heat lost during cooling is not transferred to the surrounding atmosphere. Rather, the cloud uses the heat energy to grow larger. This explains the massive structural shapes of cumulonimbus storm clouds. The rising, cooling air cannot retain as much moisture as it otherwise could if it were warm. The cloud then becomes supersaturated in moisture. Ice crystals are nucleated, but not fast enough to consume all the supersaturation. More up-drafting air entering the cloud formation creates more supersaturation and if conditions are 'appropriate', more ice crystals will form.

The 'appropriate' conditions required by the WBF Model hinge around *vapor pressure*. An easy way to envision vapor pressure is to imagine molecules of water or ice vibrating at high frequency. If one were to look at a molecule of water at extremely high magnification under a high-power microscope, the water molecule would be seen to be vibrating. This is called *Brownian motion*, named after Scottish scientist Robert Brown who coined the phrase in 1827. A molecule of water gaining enough vibratory energy (vapor pressure) will eventually turn into a molecule of steam. This transition is called a *phase change*. In the case of water, the phase change occurs at 100°C, which is the boiling point of liquid water. In the case of ice, the phase change from solid to liquid occurs at 0°C, which is the melting point of solid water.

In a cumulus or cumulonimbus cloud formation, the mass of air in the cloud will have a certain vapor pressure. This pressure V(air) will be different depending on whether it is measured at the surface of an ice crystal or at the surface of a water droplet. An ice crystal will have a unique vapor pressure V(ice) and a water droplet will have its own unique vapor pressure V(water). Three possibilities can then arise according to the WBF Model:

- If the value of a cloud's vapor pressure *V(air)* is greater than the vapor pressure of water, *V(water)*, which in turn is greater than the vapor pressure of ice, V(ice), and if the temperature is less than 0°C, the water droplets and the ice particles will both grow in size simultaneously.

- If the value of the cloud's vapor pressure V(air) is less than V(ice), which in turn is less than V(water), and if the temperature is less than 0°C, the water droplets will evaporate.

- If the value of the cloud's vapor pressure V(air) is between that of V(ice) and V(water), and if the temperature is less than 0°C, the water droplets will turn into ice particles.[2] Ice particles that form have the potential to eventually fall to the ground in the form of hail stones.

F.H. Ludlum

In the early 1960s, the focus of research shifted to the thermal updrafts in a cloud formation and the association of these updrafts to hail formation.

In a 1966 paper, *Cumulus and Cumulonimbus Convection*, F.H. Ludlum, a renowned hail researcher at Imperial College in London, UK, described up-drafting convection and cloud formation. His observations are also documented in the 1963 book *Severe Local Storms.*[3][4]

He described that a cumulonimbus storm cloud begins as a cumulus cloud without precipitation content. Over time, a series of merging, rising, thermal air masses leads to the cumulonimbus formation developing. He coined the phrase *small scale convection* to describe these rising thermal air masses.

He described how the moisture in the small scale convection air masses condenses in the growing cloud formation. The corresponding release of latent heat energy powers stronger updraft motions and grows the cloud formation further.

When the updraft speeds measure a few meters per second, the cloud formation will express raindrops. When the updrafts measure at least tens of meters per second, the result will be hailstones. As the rising updrafts subside, the cloud formation will dissipate.

If the cloud formation delivers rain, there will be an associated downdraft of air coming out of the cloud. The downdraft will hit the ground, spread out, and cause ground level, warmer air to rise upwards to further feed the cloud formation. This circulating process can persist for many hours and can cause the cloud formation to travel great distances.

If the thermal rising action persists at sufficient speed, cloud buoyancy will be increased to the point where the top of the cloud formation will be pushed high enough into the troposphere such that dense, ice-bearing particles will manifest. The presence of ice particles links back to the WBF Model. The ice-bearing particles contribute to the cumulonimbus cloud displaying its signature anvil-top formation. In the presence of wind shear, updraft speeds of 20 meters/sec (44 mph) to 40 meters/sec (89 mph) can result inside the cloud formation. This wind shear can cause ice particle growth.

Humidity, Temperature, Air Impurities, Supercooling, and Hail Formation

An expanded understanding of the WBF Model and of Ludlum's work requires a consideration of four additional factors: *humidity*, *temperature*, *air impurities*, and *supercooling*.

Humidity is the moisture content of the air in the lower atmosphere. The amount of moisture that a mass of air can hold is a function of temperature. Warm air can hold more humidity than cold air.

When a mass of air holds the maximum amount of humidity it can possibly hold at a given temperature, that mass of air is said to be *saturated*.

A mass of warm air can take in a considerable amount of moisture before it reaches full saturation. A similar sized mass of cooler air will take in less moisture before it reaches full saturation. The temperature at which a mass of air attains full saturation is termed the *Dew Point*. For example, on a given day, suppose the weather report on the radio tells you the Dew Point is 12°C. Suppose that overnight, the temperature drops to 11°C. The air would have been saturated with moisture at 12°C. Because the temperature dropped to 11°C, the air is over-saturated and can no longer hold all the moisture. The result is a covering of dew water droplets on the ground early the next morning.

Relative Humidity is the measured amount of humidity present in the air expressed as a percentage of the amount of humidity that would be present if the air were fully saturated. [5)(1]

In southern Saskatchewan where I live, a day when the temperature is 30°C can be quite comfortable because the Relative Humidity is near 40%. If I travel to southern Ontario, a day where the temperature is 30°C can feel sticky and uncomfortable because the Relative Humidity is near 90%.

In a cumulonimbus storm cloud formation, it is also possible for the Dew Point saturation and moisture condensation scenario to evolve. If the temperature happens to be below freezing when the condensation occurs, the moisture will take the form of ice crystals.

The atmosphere contains a vast array of impurities such as dust particles, smoke particles, and pollen particulates. Consider the environmental events that have made news headlines over the past several years: South American rainforests being burned by farmers, huge volcanic eruptions in places like Iceland, massive forest fires in Australia and the western US. All of these events have pushed particulates of varying composition into the atmosphere. (6)

These tiny particulates act as starting points (nucleation sites) for ice crystal formation. Formation of ice crystals at a nucleation site is not a uniform process. The type of nucleation site, the size of nucleated water droplets, and the temperature of nucleation all must be taken into consideration.

Different types of particles will initiate the freezing creation of ice at different temperatures. The notion of water freezing at different temperatures is generally not intuitive. Although we are taught in high school chemistry class that water freezes at 0°C (32°F), the controlling factor for water freezing in a cloud formation is the size of the water droplets. Large droplets will freeze at a temperature at or just below 0°C (32°F). Smaller droplets (less than 1 mm in size) can stay in liquid form down to minus 40°C (minus 40°F). These droplets are said to be *supercooled*. (1) The supercooled portion of a cumulonimbus cloud formation is the middle portion of the cloud. The lower portion is saturated with moisture and is termed the *water region*. The upper part of the formation is termed the *ice crystal* region.

What causes the temperature inside the cloud to be as low as minus 40°C? The answer rests with the fact that air cools as it rises. The rate of cooling of a mass of rising air is called the *lapse rate*. Aviation textbooks used for pilot training reference a lapse rate of 1.98°C per 1000 feet (300 meters). For example, if the temperature on the ground is 34°C, the temperature at the 35,000 foot level (10,000 meters) in the midst of a towering cumulus cloud formation with an anvil top could be as low as minus 35°C (-31°F). Small droplets of moisture (less than 1 mm in size) in the cloud can then become supercooled. Some airplanes are equipped with small TV monitors on the seat backs. The next time you are on such a flight, find the channel on the TV that shows the flight pathway. Take note of the plane's altitude, speed and more importantly the outside air temperature. A quick mathematical calculation will tell you the air cooling lapse rate.

Growth of Hailstones: Dry Growth and Wet Growth Models

Ludlum advanced two models for hail stone growth: *dry growth* and *wet growth*. Dry growth infers that the outer surface of the hailstone is composed of ice. Wet growth infers that the outer surface of a hailstone is moist. Subsequent to Ludlum's work, an intermediate model of spongy (soft hail) growth was advanced that had the surface of the hail stone comprised of alternating layers of ice and liquid.

Ludlum's growth models were based on a heat-energy balance. As supercooled moisture contacts the surface of a hailstone, the moisture releases some heat energy which causes a small rise in temperature at the surface of the stone. But, the temperature rise might not be enough to cause the surface of the stone to turn to liquid. Moisture contacting the stone will continue to accumulate on the surface and the stone will grow via dry growth. Dry growth is typically a rapid process. Air molecules often get trapped in the growing surface layers and the hailstones will take on a cloudy, opaque appearance. As the stones start to descend from the cloud formation, the thickness of the moist layer at the base of the cumulonimbus cloud plays a key role. If the moist layer is thin, the falling stone will not accumulate much additional growth. The surface of the stones will remain opaque. This is referred to as *soft hail*. However, the term 'soft' does not mean less damaging. In September 2021, I witnessed the effects of a soft hail event in west-central Saskatchewan. Although the wheat fields and canola fields had few broken stems or branches, the wheat heads were badly shelled out and the canola pods were completely slivered in half.

If the heat energy released from supercooled moisture contacting a hail stone is enough to cause the surface of the stone to turn to liquid, the surface of the stone will become sticky. Other tiny ice particles situated nearby will be attracted to the sticky surface. A hailstone developed by this wet growth mode might have a translucent appearance and an irregular shape owing to small particles that have stuck to its surface. A wet-growth stone will also have a higher drag coefficient, meaning the stone will have a slower trajectory through the cloud formation, giving more time for growth enlargement. [7]

Updraft Dynamics

Ludlum also explored the relation between thermal updraft speed in the cloud, and the *fall speed* of the hailstone.

The term fall speed refers to the terminal velocity of the falling hailstone. Fall speed is given by the formula:

$$Vt = \sqrt{(2mg)/\rho AC}$$

In this expression, m is the mass of the stone, g is the gravitational constant ($9.81 m/sec^2$), ρ is the density of the air in the cloud formation, A is the cross-sectional area of the stone, and C is the drag coefficient.

Ludlum postulated that a tiny embryo in an updraft of moderate speed will rise and accumulate growth layers and mass until its fall speed equals that of the updraft speed.

For faster updraft speeds of 25 meters/second or more, a hailstone can grow until its fall speed is twice the updraft speed.

Ludlum seems to have been motivated to explore these dynamics based on the 1937 model by Bilman and Relf that proposed that for a fall speed of 70 meters/second, a spherical stone of density 0.9 grams/cm^3 could grow to a maximum of 11 cm in diameter.[8] Personally, I hope I am never on the receiving end of an 11 cm diameter hailstone.

Ludlum believed that hailstone growth is related to repeated exposure of a small stone to the supercooled region of the cloud. He suggested an updraft event composed of a succession of thermal updrafts would allow for this repeated exposure. Ludlum also suggested that in the case of updrafts approaching 50 meters/second, hail stones can be carried aloft and rejected towards the front of the cloud's anvil formation. The falling stones can then be taken up by the rising thermals and passed through the supercooled region again and again. He based this on data from an observed hailstorm in 1961 in Wokingham, UK. This data also led him to propose that an ice crystal embryo of a certain critical size will form the largest stones in a cloud formation where the water concentration is 3 to 7 grams per m³. [3][4]

Size and Shape Distribution

In the mid-1970s, the focus of hail research turned to quantifying hailstone size. In 1977, Alberta-based weather scientists Renick and Maxwell created a nomograph model [9] (as shown in Figure 2-2) which characterizes the sizes of hailstones as ranging from shot size to larger than golf-ball size. Shot size is defined as being 1-3 mm, pea size is 4-12 mm, grape size is 13-20 mm, walnut size is 21-30 mm and golf ball size is 33-52 mm. As can be seen in Figure 2-2, an air temperature of minus 20°C at the point of maximum updraft (the point where the hailstone can no longer remain aloft), an updraft velocity of 15 meters/sec will deliver pea size hail, 20 meters/sec will give grape size hail, 30 meters/sec will give walnut sized hail. To deliver golf-ball sized hail, the storm cell would have to have an air temperature of minus 35°C and an updraft velocity of 35 meters/sec.

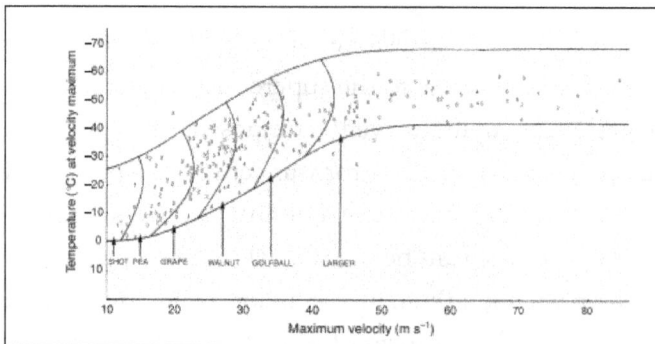

Figure 2-2
Renick & Maxwell Nomograph

The shape of a hailstone as a factor in time of formation and fall was re-visited in the mid-1980s. A 1986 study of over 6200 hailstones collected immediately after storms in Alberta and Oklahoma revealed that hailstones are not perfectly spherical. Small stones of diameter 5 mm (1/4 inch) had a long axis to short axis ratio of 0.95, and were almost spherical. Larger hailstones with diameters approaching 50 mm (2 inches) had a long axis to short axis ratio of 0.60.

Hail research in the 1980s also shifted towards embracing technology. In 1982, American weather researcher G.B. Foote studied a hail storm in Colorado. Doppler radar data was gathered along with data from an airplane that flew passes through the storm formation. Examining this data in the context of data from other researchers over the preceding decade led Foote to focus on the nature of the thermal updrafts within the storm cloud. He determined that the 1982 event had thermal updrafts that were both vertical and sloping. He further observed that the main cumulonimbus storm cloud was flanked by smaller cumulus cloud bodies containing small-scale updrafts. Researchers prior to this 1982 study had mainly assumed the existence of only one main updraft within which hailstone growth occurred. Based on data from the 1982 storm, Foote was able to advance the idea that hailstones can grow in a flanking cumulus cloud formation. As the stones then move into the main updraft zone in the cumulonimbus formation, the stones will grow further in size. But, Foote also conceded that not every hail event will come from a formation that resembles the 1982 Colorado event. In fact, he noted that no two hailstorms will be of the same size or same composition.

Foote used the 1982 data to formulate a model based on: a thermal updraft with a finite lifetime, a given diameter, and a horizontal wind flow. His model states:

$$t\alpha = dU^{-1}$$

In this model, $t\alpha$ is the time scale for hailstone movement and growth, d is the thermal updraft diameter, U is the horizontal wind flow.

In practical terms, as the updraft diameter increases (larger storm cloud formation), the time scale for stone growth increases and larger stones are possible. For a given updraft and storm cloud, if the horizontal wind flow diminishes, the time scale for stone growth increases and larger stones are possible.

Foote also introduced the variables tm and τ, where tm is the time needed for a stone to grow and τ is the lifetime duration of the updraft in the cloud.

In practical terms, if tm is greater than τ, the growth of stones is limited to the lifetime span of the updraft. If the time needed for stone growth (tm) is greater than $t\alpha$, there will not be enough time for the stones to realize their full growth potential. This will be the case with a narrow updraft (d) or a large horizontal wind flow (U). These two observations go a long way towards explaining why not every storm cloud delivers hailstones in a uniform predictable manner. Foote further stated that if τ is greater than $t\alpha$ which is greater than tm, then conditions exist for a significant hail storm event. [10]

Future Expectations

What can we expect from hailstorms in the future? The short answer is – nobody knows with certainty. Predictive models are only as good as the data input into the models.

Past statistics provide average occurrence numbers. But these numbers are not uniform. Embedded within the past occurrence data are years with elevated storm numbers. The number of hailstorm events is one matter. The severity of the events is another matter.

Based on past data from 1982 to 2006, it seems reasonable to expect just over 40 hail events in a growing season in Saskatchewan. [11] In Alberta, data from the *Alberta Hail Project* suggests to expect a similar number of hail events with many of them being severe. In fact, the city of Calgary and surrounding area has an unsavory reputation for its devastating hail events. [12] According to the NOAA Storm Database, Wyoming, Texas, and Oklahoma might experience up to 48 hail events per year. Nebraska and Colorado might experience closer to 70 events per year. [13]

As computing technology has increased, so too has academic effort to model and predict hailstorms using output from weather prediction models. Models of all sorts, however complex they may appear, are generally based on just a handful of assumptions. Weather models in North America largely center around The North American Climate Change Assessment Program (NARCCAP). This is an atmospheric-ocean circulation model that can be run based on a number of scenarios. One scenario used in hail modelling is the A2 scenario which assumes by 2050 the world will have: a global population of 10 billion people, annual CO_2 emissions equivalent to 600 Giga-tonnes of elemental carbon, atmospheric CO_2 levels of 575 parts per million (ppm), and SO_2 emissions equivalent to 105 Mega-tonnes of elemental Sulfur. These assumptions appear reasonable, but should not be taken as a firm guarantee of future emissions levels.

In 2017, a group of Canadian scientists fed data from the A2 NARCCAP model into a hail model program called HAILCAST. Attempting to model hailstorms is difficult and fraught with variables including: how do the hail stones tumble as they fall, to what extent is moisture retained on the stone surface, what is the density of the stones, and so on. The HAILCAST model begins with a liquid droplet at the base of a storm cloud that has been predicted to occur by the NARCCAP model. The liquid embryo droplet (300 μm in size, or 1/3 of a mm) is assumed to experience an updraft of 4 meters/second. When the droplet encounters the layer of the storm cloud that is minus 8°C, the droplet freezes. The frozen droplet is assumed to contain some air giving the droplet a density of 900 kg/m³. The droplet continues to grow so long as the temperature in the cloud is between minus 20°C and minus 40°C.[14][15] The HAILCAST was run using data from 1971 through to 2000. This allowed the scientists to compare model output to actual documented hail events. Next, the model was run using future predicted NARCCAP storms for a period 2041-2070. To keep the amount of model data manageable, the HAILCAST model was only run for time periods March 1 to September 30, and only for days when surface temperature was >10°C.

This HAILCAST study output aligns to the work of Lei Wang at Harvard. Wang's research points to an increase in hail severity in areas of western Canada and the US High Plains. This is the same geographic area where Wang suggests hotter and drier conditions will develop. The implications for agricultural crops are significant. Future growing seasons could become very dramatic for farm operators. Hail insurance and crop insurance are input costs that cannot be ignored. Farm operators may have to consider maximizing both types of insurance.

To Sum it Up

» A faster-moving warm air mass rising over top of a slower-moving cold air mass creates a *warm front*.

» A cold air mass bumping against a warm air mass creates a *cold front*.

» Cumulonimbus clouds are formed from cold fronts with help from thermal updrafts. These clouds are also called *storm clouds* or *storm cells* and are often the origin of hail storm events.

» In the 1930s, European scientists Wegener, Bergeron, and Findreisen developed the WBF Model of storm cloud formation. The WBF Model is based on the 1st Law of Thermodynamics and adiabatic cooling of rising air masses.

» The WBF Model is focused on the vapor pressure of the air mass in the cloud formation and the vapor pressure of moisture in the cloud. If conditions are right, ice crystals can form which can fall to the ground in the form of hail stones.

» In the 1960s, British scientist F.H. Ludlum suggested that ice crystals in a cloud formation can grow by dry growth or wet growth. He proposed that hailstones in cloud formations will grow until their fall speed equals the thermal air updraft speed. He further proposed that hailstone growth was related to repeated exposure of a small stone to the supercooled region of the cloud.

» In the 1970s, Alberta researchers Renick and Maxwell characterized the sizes of hailstones as ranging from shot size to larger than golf-ball size. Shot size is 1-3 mm, pea size is 4-12 mm, grape size is 13-20 mm, walnut size is 21-30 mm and golf ball size is 33-52 mm.

» In 1982, American weather scientist G.B. Foote created a model based on a thermal updraft with a finite lifetime, a given diameter, and a horizontal wind flow. He further added to the model a time factor needed for stone growth and a duration factor for the thermal updraft in the cloud.

» Modern modelling uses the A2 NARCCAP scenario data fed into models such as HAILCAST. Modelling is suggesting an increase in hailstorm severity going forward.

» Hail insurance and crop insurance will critical items to purchase in the future.

Frozen Fury

CHAPTER 3
CAN HAIL BE SUPPRESSED?

As the scientific understanding of hail formation has deepened over the years, the question that has repeatedly arisen is - can a developing cumulonimbus cloud be disrupted from reaching the point where it delivers a damaging hailstorm? The short answer to this question is - maybe.

Vonnegut and Schaefer

If the surname Vonnegut sounds familiar, it might be because you have read *Cat's Cradle*, written in 1963 by American author and satirist Kurt Vonnegut.[1] In this novel, Vonnegut describes a fictional substance called *ice-nine* which when added to liquid water causes the water to turn to ice. In the early 1940s, Vonnegut studied chemistry at Cornell University and then anthropology at the University of Chicago. Neither academic pursuit proved successful. From 1947 to 1950, he managed to hold a job at General Electric as a technical writer. His dream, however, was to become a published novelist. After successfully getting paid to write a number of magazine articles, he left General Electric to pursue novel writing full time.

Vonnegut's brother, Bernard, may have been the motivation for the fictional substance called *ice nine*. Bernard Vonnegut was more educationally accomplished than his brother. Armed with a pH.D. in physical chemistry from M.I.T. in Boston, Bernard went to work for General Electric in 1945 as a research scientist. Prior to arriving at General Electric, Bernard had studied airplane-wing ice formation and strategies to reduce it. Once at General Electric, Bernard collaborated with Vincent Schaefer who was studying the manufacture of snow particles by exposing super-cooled water to a small atmospheric particle of dust or salt. Vonnegut started studying the use of silver iodide crystals as an alternative to dust and salt particles. He successfully demonstrated the use of silver iodide crystals as nucleation sites to cause super-cooled water to form ice particles. Did this discovery motivate Kurt Vonnegut to write about the fictional substance *ice-nine* in *Cat's Cradle*? It would appear so.

Schaefer and Vonnegut went on to collaborate on the use of dry ice particles to seed actual cloud formations. On November 13, 1946, six pounds of dry ice particulate was seeded into a cloud formation in upstate New York. The result was a significant dump of snow in western Massachusetts. A second cloud seeding effort created a major snowstorm on December 20, 1946. Lawsuit rumblings from the public caused General Electric to pull back from its cloud seeding research.

Vonnegut turned his attention to other research projects at General Electric. Schaefer went to work for the US military on Project Cirrus, a weather modification program designed to weaken hurricanes. On October 13, 1947, an airplane seeded 180 pounds of dry ice particulate into a hurricane that was heading out to sea. The storm suddenly reversed course and made landfall at Savannah, Georgia, causing significant damage. Project Cirrus was abruptly cancelled as threats of lawsuits mounted.

Schaefer eventually went to work for the U.S. Forest Service in northern Idaho to determine the effect of cloud seeding on the patterns of lightning in thunderstorms.

Vonnegut eventually left General Electric and took up a teaching post at the University of Albany. He remains the scientist of record who discovered the use of silver iodide crystals to seed cloud formations so as to cause ice and snow crystals to form. (2)

Soviet Hail Experiments

As Vonnegut and Schaefer were conducting their work at General Electric, half-way around the world Soviet scientists were also investigating cloud seeding as a means of possible hail suppression. In 1964, a group of seven American climate scientists was invited to Russia to meet their Russian colleagues. On one hand, the visiting delegation was surprised at the level of science being pursued by the Russians. On the other hand, the meetings proved less illuminating than expected.

Soviet researchers Levin and Litvinov at the Institute for Geophysics revealed that cloud seeding experiments had been ongoing for many years. However, the results of attempting to modify snowstorms had not demonstrated significant progress. At the Geophysical Observatory in Leningrad, Dr. Nikandrov showed the visiting delegation the apparatus for shooting artillery shells containing silver iodide particles into convective clouds. No statistically significant results had been obtained. At the Kiev Hydrometeorological Scientific Research Institute, Dr. Prekhochko described his efforts to use dry ice to seed storm clouds. No results of any statistical significance had been obtained. At the Alpine Research Institute, with headquarters in Nalchik, Dr. G. K. Sulakvelidze described efforts to launch balloons filled with silver iodide into cumulonimbus clouds. He reported that 80 balloons out of 200 had generated some hail. Dr. G. K. Sulakvelidze introduced the visiting delegation to his hail model which posits that hail is produced as follows:

- Large supercooled water droplets freeze and form hailstone embryos.

- The embryos grow very rapidly by coalescence in the region of the cloud containing of high liquid water content.

- The hailstone embryos are prevented from falling out of the cloud by strong thermal updrafts.

The visiting delegation noted that Sulakvelidze's model closely resembled the model of F.H. Ludlum. Sulakvelidze stressed that his model differed from that of Ludlam's hailstorm model in that Ludlum assumes a tilted updraft comprised of rising thermals and wind shear. Sulakvelidze noted that his model assumed a nearly vertical thermal updraft.

Sulakvelidze went on to describe his efforts at launching silver iodide particles (12 μm diameter) into clouds by way of artillery shells. He pointed out that his work showed each gram of silver iodide produces 1×10^{13} to 1×10^{14} nuclei sites and that one hail particle was produced for every 5000 to 10,000 silver iodide particles. Despite this numerical insight, Sulakvelidze admitted that in no case in which silver iodide was fired into clouds over the target area did significant amounts of hail drop out of the clouds. Perhaps to cover his lack of practical progress in seeding clouds, Sulakvelidze also claimed to be doing work using sound waves directed at cloud formations as a means of altering the formation of hail.

At the Georgian Academy of Science in Tbilisi, Dr. A. I. Kartsivadze described work incorporating 75 gram quantities of silver iodide blended into rocket fuel. The fuel was used to propel small rockets into cumulonimbus cloud formations with an estimated internal temperature of about minus 10°C. As the fuel burned, an exhaust stream of silver iodide particulate was emitted that produced an estimated 5×10^{12} nuclei per gram of silver iodide. Kartsivadze described how his rocket efforts had been done over an area of 80,000 hectares. In the 1962 hail season there were only three hailstorms over these 80,000 hectares while in a nearby control area there were 13. During 1963 there were 120,000 hectares in the target area. Hail in this area occurred 5 times while in a nearby control area it occurred 14 times. He eventually admitted that data for the 25 preceding years showed that in the cloud-seeded test area there were an average of about 6 incidents of hail while in the control area there were 5.8 cases of hail. Despite this apparent statistical insignificance, he was adamant that his work was effective at modifying storm clouds. The visiting delegation of American scientists left Russia satisfied that Soviet scientists had not made any major breakthroughs in cloud seeding. [3]

North Dakota Cloud Modification Program

One of the oldest cloud seeding programs in operation is the North Dakota Cloud Modification Project (NDCMP), which has been in existence since 1951. This program was likely influenced by the earlier work of Schaefer and Vonnegut at General Electric. The NDCMP project in North Dakota is conducted using airborne-released particles of silver iodide (or dry ice) into convective clouds thought to be have hail-producing properties. The silver iodide crystals transform supercooled cloud moisture droplets within the cloud into large numbers of tiny ice crystals. In doing so, large hailstone growth is reduced, thus rendering any falling hail less destructive. A 2020 study of past data (1989-2018) found that across 1490 data points the NDCMP cloud seeding program improved wheat yields by just over 3 bushels per acre in the counties of North Dakota where cloud seeding was conducted. This yield improvement in absolute terms is obviously small and the statistical goodness of fit (R^2) value for regression studies described in the analysis was less than 0.60. In other words, the effectiveness of cloud seeding in North Dakota is questionable. [4]

Cloud seeding to modify hail events is based on the concept of *beneficial competition*. Beneficial competition assumes there is a deficiency of tiny nucleation sites in the cloud formation. It assumes the injection of silver iodide will produce a significant number of ice nucleation sites. The existing natural nuclei sites and the artificial ice crystals will compete for the available supercooled liquid cloud water within the cloud. As a result, the hailstones that are formed within the seeded cloud will be smaller and will generate less damage if they should survive the fall to the surface without melting. (5)

Alberta Hail Project

Another long-running cloud modification program is located in Alberta. In 1996, Weather Modification Inc., the same contractor who conducts the annual work in North Dakota, was hired for a five-year project by the Alberta Severe Weather Management Society (funded by a consortium of property and casualty insurance underwriters). The contracted task was to conduct cloud seeding to reduce urban property damage from hail, particularly in the Calgary and Red Deer areas. This program has since been extended beyond its initial five year mandate, and continues to operate to this day. Three aircraft are situated at the Springbank airport just west of Calgary and two aircraft are at the Olds airport north of Calgary. If a potentially dangerous cloud formation is identified by radar, planes are mobilized to disperse silver iodide crystals into various levels of the cloud formation. It is estimated that this hail suppression effort has reduced damages to crops by 20%. However, climate change, effectiveness of cloud seeding on particular clouds, and genetic improvements to crops (ie shatter resistant canola) are all variables that make hard and fast conclusions difficult. The World Meteorological Organization (WMO) summed up the argument in 2015 when it stated that: *glaciogenic seeding technologies have been used operationally in many parts of the world to reduce hail damage. However, scientific evidence to date is inconclusive and evaluation of the results has proven difficult.* (6)

Argentina

The province of Mendoza in Argentina has been the location of hail suppression projects for many years. Wine drinkers might recognize the name Mendoza. This area of Argentina produces 75% of the world's Malbec wine. Owing to its geographic location near the Andean foothills, this area is subject to significant hail events each year that destroy around 10% of the grape crops. Hail suppression projects in the Mendoza area between 1959 and 1964 were concluded to have been ineffective. Efforts between 1979 and 1982 were also labelled as ineffective. Efforts are ongoing, but more evidence is needed before success can be claimed. [7]

Hail Suppression Remains A Complex Subject

There are a number of interwoven variables that influence the ability of a cloud formation to deliver hail: moisture content, temperature at the cloud base, degree of wind shear in the cloud formation, strength of the up-drafting wind, temperature of the super-cooled region in the cloud, the number of nuclei sites in the cloud, the number of silver iodide nucleation sites seeded into the cloud, and atmospheric pressure. With such complexity, it is easy to see why the answer to the hail suppression question is – maybe.

To Sum It Up

» In the 1940s, Bernard Vonnegut at General Electric used silver iodide particles to seed cloud formations.

» In the 1950s, Soviet researchers conducted a number of cloud seeding programs. None demonstrated significant efficacy.

» The State of North Dakota has an ongoing cloud seeding program. Results point to a marginal statistical benefit at best.

» An ongoing cloud seeding program in the Calgary and Red Deer areas of Alberta indicates a beneficial outcome, but statistical quantification is difficult to calculate.

» A multi-year cloud seeding in effort in the wine-producing region of Argentina showed inconclusive results.

» The answer to the question – can hail storms can be suppressed? remains a definite - maybe.

Frozen Fury

CHAPTER 4
SEED STRUCTURE AND PLANT GROWTH

Having taken a look at weather, climate, and the physics of hail, let's now shift our focus over the next few chapters to a look at: seed structure and the biology of plant growth of several of the crop types that grow in western Canada and the US northern Great Plains. These are the crops that are often damaged by hailstorms each year.

To better appreciate agricultural plants and their growth one must go back in time to the end of the last Ice Age, some 20,000 years ago, when a massive global warming event brought an end to the sheet of ice that covered much of the planet.

Pleistocene to Holocene

Geologists refer to the period of the last Ice Age as the *Pleistocene Epoch*. The last Ice Age reached its peak about 20,000 years ago according to consensus among academic climate researchers. Following this peak, the ice started to melt. As the ice receded, the global climate shifted from being cool and dry to being warmer and moister. The post-Ice Age period is referred to as the *Holocene Epoch*. As the ice melted and the climate warmed, primitive Neolithic man began to evolve and adapt to his new circumstances.

As the planet warmed, Mother Nature evolved too. New species of plants began to take form. Mankind learned over time that the seeds of these plants could provide nutritional sustenance. To collect these seeds, early man learned that each autumn it was necessary to walk considerable distances to collect the ripened seeds of plants. Man became a skilled gatherer.

Younger Dryas

But the transition to warmer temperatures would encounter an obstacle that would present Neolithic man with a challenge. Starting about 13,000 years ago, and lasting for about 1300 years, the global climate shifted back to being cooler and drier. This event is what researchers call the *Younger Dryas* period. The exact reasons for this shift remain unclear.

Agriculture is Domesticated

The Younger Dryas period is responsible for modern day agriculture as we know it. As the climate became cooler and dryer, the proliferation of plants and seed-bearing wild grasses diminished. Neolithic man came to the realization that to survive he could not rely on walking around vast areas of territory in search of seeds. Early man noticed that grasses yielded seed-bearing heads that easily were shattered and blown into the wind when ripe. But a small percentage of these grasses had more durable heads that prevented wind-borne seed losses. Motivated by time and labor, early man learned to collect the seeds from these tougher plants. The seeds were then planted in a central location. The effort required to obtain food was thus reduced. Mankind also learned to keep animals in close proximity. This reduced the need for walking vast distances in search of game. The Younger Dryas period diminished the need for hunting and gathering and shifted early man onto a path towards a domesticated lifestyle model. Our lifestyles to this day are domesticated and sedentary all thanks to the Younger Dryas period that challenged our Neolithic forebearers to domesticate seed-bearing grasses. [1]

Plant Categories

Broadly speaking, seeds can be divided into two categories: monocotyledons (*monocots*) and dicotyledons (*dicots*). Monocots differ from dicots in four aspects: roots, stems, leaves, and flowers.

Monocots generally have a fibrous root system that branches off in many directions. The root system tends to occupy the upper levels of the soil. The sketch in Figure 4-1 illustrates the monocot root system.

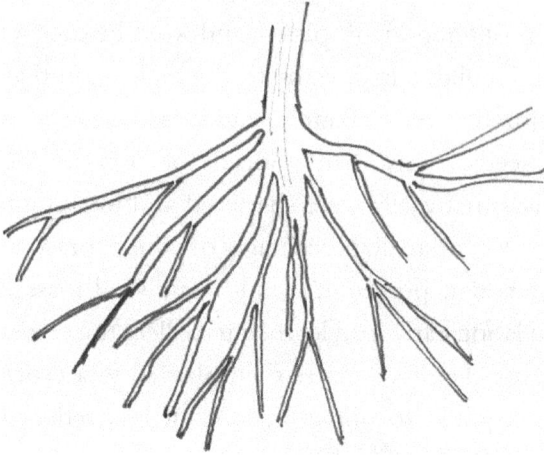

Figure 4-1
Monocot Root System

Dicots have a thicker, denser root system built around a tap root (main root). The sketch in Figure 4-2 illustrates the dicot root system.

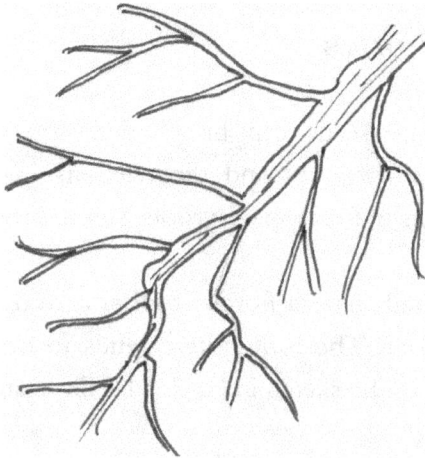

Figure 4-2
Dicot Root System

Inside a plant stem there is a vascular system (circulatory system) to bring moisture and nutrients from the soil into the roots, and then into the plant tissue. Monocots have their vascular system arranged in a sporadic fashion. Dicots have a more orderly arrangement. The cross-section sketches in Figures 4-3 and 4-4 illustrate.

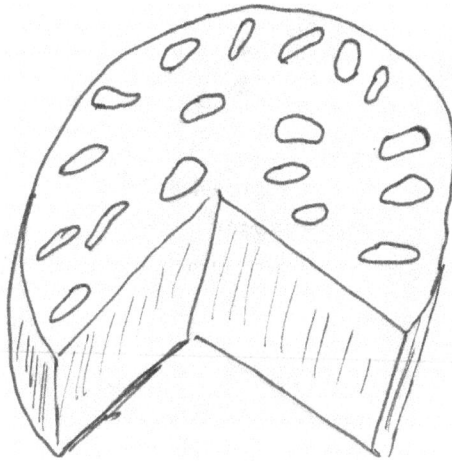

Figure 4-3
Cross section of a monocot stem
showing the sporadically arranged vascular system

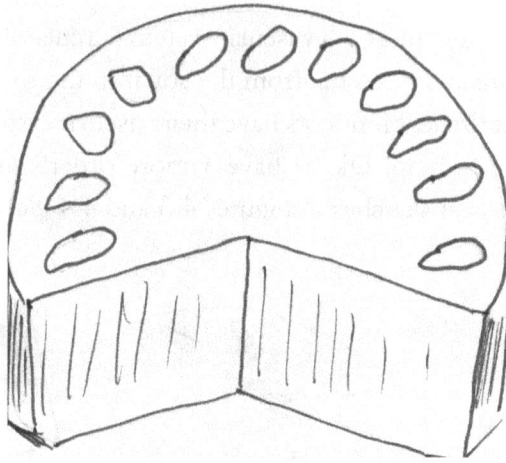

Figure 4-4
Cross section of a dicot stem
showing a more orderly arranged vascular system

To the casual observer, an easy way to distinguish monocots from dicots is by studying the leaves. Monocot leaves have parallel veins. Dicot leaves have branching veins. Figures 4-5 and 4-6 illustrate.

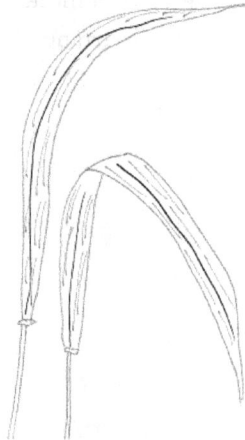

Figure 4-5
Monocot leaf structure with parallel veins

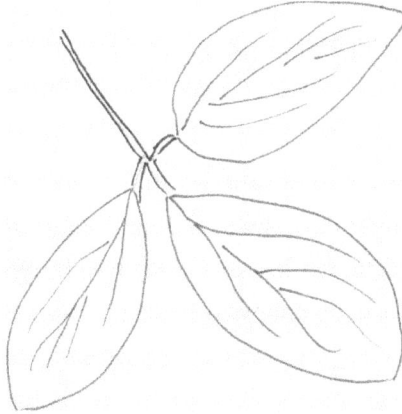

Figure 4-6
Dicot leaf structure with branching veins

Lastly, if the plants bear flowers, a monocot plant will express flower leaves in multiples of three. A flowering dicot plant will express leaves in multiples of four or five. For example, a poppy flower is a dicot and displays four or five leaves. A tulip flower with three outer leaves is a monocot. (2)

Seed Structure

Broadly stated, there are three components to a seed: the *seed coat*, the *endosperm*, and the *embryo* (consisting of epicotyl, hypocotyl, radicle, and cotyledon tissues). Mother Nature has a handful of molecular structures at her disposal (lignins, lipids, proteins, and carbohydrates) and she uses them in very creative ways to create the seed structures for the food substances that sustain life for humanity.

Seed coat cells are comprised of a complex mix of *lignins* and *lipids*.

Lignin molecules are a complex array of carbon, hydrogen, and oxygen atoms. Lignin is responsible for the mechanical strength of the seed coat.

Lipids are present throughout the world of biology and microbiology. A lipid consists of a 3-carbon atom backbone called glycerol. Attached to this backbone are two fatty acids (a fatty acid is a chain of carbon atoms with hydrogen atoms attached) and a phosphate residue (a phosphorous atom with four oxygen atoms attached). The basic lipid structure lends itself to the formation of everything from waxes to cholesterols. In the seed coat, lipids play a complex role in helping the seed to germinate.

Endosperm cells are situated within the body of a seed. An endosperm cell is comprised of 3-layers: an inner layer of protein and two external layers comprised of the carbohydrates *arabinoxylan* and *beta glucan*. Arabinoxylan is a *hemi-cellulose* structured molecule. A hemicellulose molecule comprises a 5-carbon structure with hydrogen atoms and hydroxy (OH) molecules attached. Beta glucan is comprised of chains of glucose molecules which are comprised of a 6-carbon backbone with hydrogen atoms and hydroxy (OH) molecules attached.

The various embryo cellular tissues (consisting of epicotyl, hypocotyl, radicle, and cotyledon tissues) are comprised of lignins, lipids, proteins, and carbohydrates.

Seed Tissues

Notable features of a seed embryo structure include the *epicotyl tissue*, the *hypocotyl tissue*, and the *radicle tissue*. The prefix *epi* means that which is above. The prefix *hypo* means that which is below. In the embryo structure, the epicotyl tissue is situated above the hypocotyl tissue. Figures 4-7 and 4-8 illustrate this. The epicotyl tissue will eventually grow to form the leaves of the plant.

In a monocot seed, the epicotyl tissue is encased in a sheath tissue called the *coleoptile*. The hypocotyl tissue will eventually give form to the lower stem of the plant. The radicle tissue will give form to the roots of the plant. In a monocot seed, the radicle tissue is located in the embryo structure enclosed in a sheath called the *coleorhiza*.

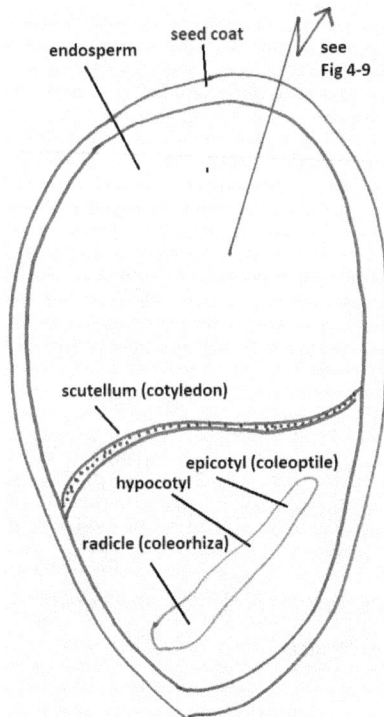

Figure 4-7
Monocot Seed in cross section

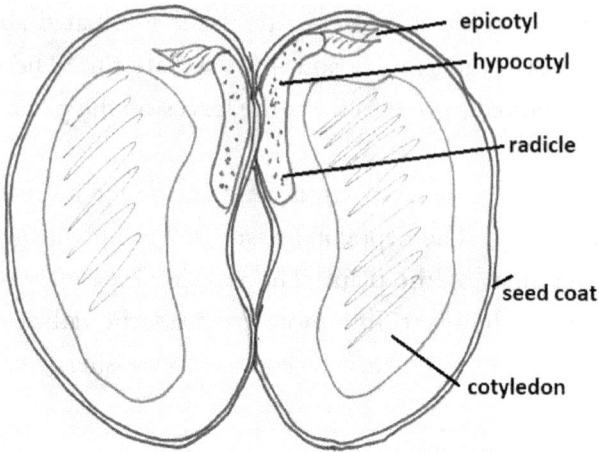

Figure 4-8
Dicot Seed in cross-section

Monocots

In monocots, the seed embryo tissue is distinguished from the endosperm cell structure by a membrane of vascular tissue called the *scutellum*. As Figure 4-7 shows, this membrane is also referred to as the *cotyledon*. It is the conduit for carbohydrate and protein energy from the endosperm cells to reach the embryo. Figure 4-9 illustrates a cross-section through an endosperm cell. In the endosperm cell the larger globules are starch (carbohydrate) and the smaller, more numerous globules are proteins.

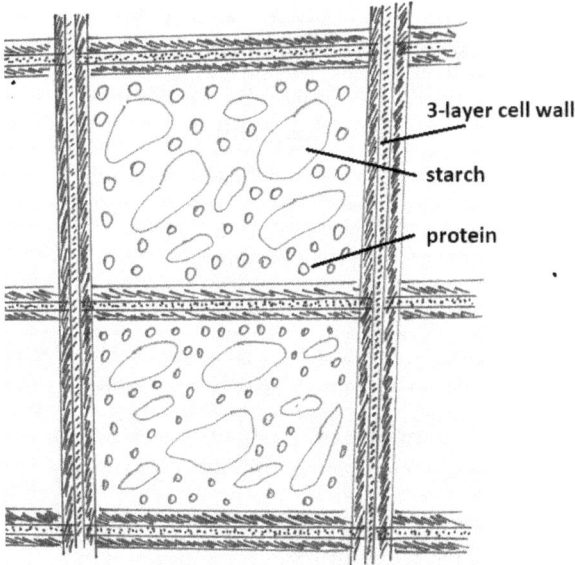

Figure 4-9
Endosperm cells in cross section

When the monocot seed is planted in the ground, moisture is taken into the seed structure through the seed *micropyle end* which is located near the embryo (a monocot seed requires at least 30 percent moisture to complete the germination process). The opposite end of the seed is the *awn end*. The moisture interacts with the scutellum tissue (cotyledon) to generate a hormone called *gibberellic acid*. This acid induces a configurational change in the protein cells adjacent to the seed coat. The altered proteins become *enzymatic* and are now capable of breaking down the 3-layer cell structure of the endosperm cells. As the endosperm cells break down, the stored starches and proteins within the endosperm cells are released and become available for the seed to use in order to commence the growth of the epicotyl, radicle, and hypocotyl tissues.

The first tissue material to emerge from the micropyle seed end during germination is the radicle, which goes on to form the root system. The next tissue to emerge is the hypocotyl which within several days will poke up through the soil and become visible as a single leaf structure and eventually the stem structure. This stem structure goes on to elongate, form additional leaves, and eventually forms a reproductive *caryopsis* or 'kernels'.

Here is a critical point. In a monocot, the cotyledon tissue material *never does* emerge above the soil. This is a key distinction between monocot plants and dicot plants. Plants in which the cotyledon tissue remains beneath the soil are said to exhibit *hypogeal germination*.

As an aside, in beer brewing, malted barley is the raw material used. The malting process is a controlled version of the same germination process that occurs when a seed is planted in a farm field. The malting company will steep the grains in water for eight hours, remove them from the water for eight hours and then repeat this process one more time. The moisture that the grains uptake through their micropyle ends is sufficient to invoke germination. Once the seeds begin to express a root and a shoot, heat is applied to halt any further development. The broken-down, partly-altered endosperm cell structure now provides the beer brewer with a source of fermentable sugars (carbohydrates) and proteins that yeast will consume to generate alcohol. [3]

Dicots

In dicots, there are two cotyledon membranes as Figure 4-8 illustrates in cross-section. As a dicot seed develops, the carbohydrate and protein energy reserves are mobilized from the endosperm structure into the cotyledons. In many dicot seeds, the amount of remaining endosperm material is minor. When a dicot seed is planted in the soil, the seed uptakes moisture which induces protein tissue to become enzymatic. Through a complex series of reactions, trace amounts of ethylene (C_2H_4) are formed which trigger the onset of tissue growth.

The first material to emerge from the seed is the radicle which goes on to form the tap root system.

Next, the hypocotyl tissue expresses growth and elongation. Here is a critical point. The epicotyl tissue does not elongate. As a result, the hypocotyl tissue pulls the cotyledon tissue up and through the soil. The cotyledon tissue quickly opens and develops into leaf material. The process of photosynthesis then provides growth energy for the plant. Plants which express cotyledon material above the soil are termed *epigeal germinators*.

Examples of Monocots and Dicots

Monocot plant examples include: cereal grains, corn, rice, canary seed, grasses, lilies, asparagus, yams, leeks, and onions. As Figure 4-5 showed, the parallel leaf veins are the distinguishing factor in monocots. In the context of hail insurance, early-stage cereal grain plants can be damaged by hailstones from an early-season storm. Because the cotyledon tissue is beneath the soil (hypogeal), the damaged plant will have enough energy to express re-growth of leaf and stem material.

Dicot plant examples include: mango, buckwheat, rhubarb, lentils, field peas, chickpeas, canola, mustard, soybeans, faba beans, flax, sunflower, hemp, cannabis, cucumber, pumpkins, tomatoes, sugar beet, gourds, French beans, watermelon, and lettuce.

In the context of hail damage, early-stage epigeal plants can be damaged by hailstones from an early-season storm. If some cotyledon tissue remains after the storm, the plant has a good chance at recovery. If the cotyledon tissue has been thoroughly damaged, the plant will not recover.

Examples of Epigeal and Hypogeal Germinators

From the above monocot and dicot plant examples, those that are hypogeal germinators include cereal grains, corn, rice, canary seed, grasses, mango, asparagus, yams, leeks, faba beans, chickpeas, lentils, and field peas.

Monocots and dicots that are epigeal germinators include: buckwheat, rhubarb, flax, French beans, sunflowers, tomatoes, cucumbers, pumpkins, watermelon, gourds, lilies, onion, mustard, hemp, cannabis, soybeans, sugar beet, and canola.

To Sum It Up

» The last Ice Age occurred in the Pleistocene Epoch.

» The end of the Ice Age marked the start of the Holocene Epoch.

» The Younger Dryas period is responsible for modern day agriculture as we know it.

» Seeds can be divided into two categories: monocotyledons (*monocots*) and dicotyledons (*dicots*).

» Monocots generally have a fibrous root system that branches off in many directions.

» Dicots have a thicker, denser root system built around a tap root.

» Monocots have the vascular system arranged in a sporadic fashion. Dicots have a more orderly arrangement.

» Monocot leaves have parallel veins. Dicot leaves have branching veins.

» A monocot plant will express flower leaves in multiples of three. A flowering dicot plant will express leaves in multiples of four or five.

» The three components of a seed are: the *seed coat*, the *endosperm*, and the *embryo*.

» Seed coat, endosperm, and embryo tissues in a seed are made of lignin, lipid, hemicellulose, arabinoxylan, glucose, beta glucan and protein structures.

» A seed embryo structure includes the *epicotyl tissue*, the *hypocotyl tissue*, and the *radicle tissue*.

» Seeds germinate via epigeal or hypogeal methods. In epigeal germination, the cotyledon tissue is pushed above the soil. In hypogeal germination, the cotyledon tissue remains beneath the soil.

» Monocot plant examples include: cereal grains, corn, rice, canary seed, grasses, lilies, asparagus, yams, leeks, and onions.

» Dicot plant examples include: mango, buckwheat, rhubarb, lentils, field peas, chickpeas, canola, mustard, soybeans, faba beans, flax, sunflower, hemp, cannabis, cucumber, pumpkins, tomatoes, sugar beet, gourds, French beans, watermelon, and lettuce.

» Epigeal germinators include: buckwheat, rhubarb, flax, French beans, sunflowers, tomatoes, cucumbers, pumpkins, watermelon, gourds, lilies, onion, mustard, hemp, cannabis, soybeans, sugar beet, and canola.

» Hypogeal germinators include cereal grains, corn, rice, canary seed, grasses, mango, asparagus, yams, leeks, faba beans, chickpeas, lentils, and field peas.

CHAPTER 5
PLANT GROWTH STAGES

Vegetative and Reproductive Growth

Plants exhibit two stages of growth: vegetative growth and reproductive growth.

The vegetative growth of a plant will see the emergence of tissue material from the seed, formation of stalks, stems, secondary branches and leaves.

Reproductive growth will see Nature propagate itself by generating seeds and fruits on the plant structure after vegetative growth is complete.

Nature expects that these seeds will ripen and fall off the plant under the influence of wind and rain, thus setting the stage for another season of growth. In Neolithic times, man learned to intervene when the plant exhibited seeds and fruits. Collected material provided sustenance until the next growing season delivered more plants and vegetative growth. Today, mankind still intervenes as plants exhibit reproductive growth, except intervention can take the form of a combine and grain storage bins.

Growth Signals

Different plants will switch from vegetative growth to reproductive growth at different times of the season and under different conditions. The precise pathway for this switch is not fully understood across all plant types. Scientists continue to make progress in their understanding of plant growth thanks to technological advances that now give the ability to identify individual gene sequences in plants.

The cells that comprise plant leaves contain *phytochrome* and *cryptochrome* protein substances that can trigger a gene response signal in the plant's metabolic pathways.

Phytochrome proteins react to changes in red and far-red light wavelengths. Red light forms part of the visible light we experience in the daytime. The visible light spectrum covers the range 400 to 700 nanometers. Red light has a wavelength of 700 nanometers. Far-red light with a wavelength of 780-800 nanometers is not visible to the human eye, but plants can detect it by way of the phytochrome proteins.

Cryptochrome proteins react to changes in blue light and UV-A light. Blue light has a wavelength of 450-495 nanometers. UV-A light with a wavelength of 315-400 nanometers is not visible to the human eye, but plants can detect it by way of the cryptochrome proteins.

Under the influence of red light, phytochrome will switch to its active form and vegetative growth will be signaled by the plant genes. As night falls, or as daytime hours grow shorter as the growing season progresses, far-red wavelengths enter the situation. Under the influence of far-red light, phytochrome switches to its inactive form and vegetative growth will diminish and eventually cease. A similar model applies to the blue and UV-A light.

Plant response to varying amounts of daylight is called *photoperiodicity*. The photoperiodic category is divided into *long-day* and *short-day* sub-categories.

Long day plants will start to flower (reproductive growth) when darkness hours are at or near a minimum (light hours are at or near a maximum). Plants like Oat, Barley, Wheat, Flax, Canola, Mustard, and Pea will commence flowering in the time immediately after June 21 (Summer Solstice) when the days are longest. Short-day plants will start to flower when dark hours exceed a minimum threshold. Rice, Cotton, Cannabis, Hemp, and Beans are examples of short- day plants. [1][2][3]

Crop History and Growth Descriptions

Flax *(Linum angustafolium)*

- Seed Structure: Dicot
- Growth: Epigeal
- Vegetative Growth: 45-60 days
- Flowering: 15-25 days
- Maturation period: 30-40 days

Flax was an early dietary crop that Neolithic man encountered as the Ice Age ended and the climate warmed. Professor Chris Cullis of Case Western Reserve University, writing in the 2007 publication *Oilseeds*, presents a comprehensive look at flax. The first archeological evidence of flax dates to 9000 years ago in what is now modern-day Turkey. Evidence dating to 8000 years ago has been found in modern-day Syria and Iran. Flax was being cultivated by early man in India and Egypt 5000 years ago. The early Egyptians learned to use flax plant fibers to make ropes and fabrics. The Egyptians crushed flax seeds manually using a stone mill to obtain flax oil for lubrication. They also exposed seeds to hot water to liberate the oil, which was then used for dietary consumption.

Archeological evidence suggests that flax was brought to western Europe as the Roman empire expanded. In 1617 Lois Hebert brought flax seed to New France (modern day Quebec, Canada). By 1875, flax was being grown in western Canada thanks to settlers who had pushed west as the railroad was constructed. Settlers from Europe coming to the American Colonies also brought flax seeds with them. The soils in the Colonies proved compatible to flax seeds. Settlers created linen materials from flax fiber as well as wood preservatives from the flax oil. The polyunsaturated flax oil with its linoleic acid content forms a thin protective film when applied to a wooden surface. The oil (called linseed oil) also eventually found its way into oil-based paints. Flax meal left over after oil extraction was used as animal feed. These trends largely continued through to World War II. The post-War era saw the introduction of synthetic fibers and with that, flax demand started to decline. In the 1980s, water-based latex paints began to replace oil-based paints and flax demand dropped further. This demand drop, however, did not persist. Through plant breeding efforts, scientists were able to modify the structure of flax plants so as to generate oil that closely matched the molecular structure of the healthy omega oils found in fish. Today, flax is primarily grown for consumption rather than for industrial use. A bushel of flax will yield approximately 2.6 gallons of flax oil. [4][5][6]

Professor Cullis at Case Western Reserve University describes 12 growth stages of flax as follows:

1. Cotyledon
2. Growing point emerges
3. First pair of true leaves unfolds
4. Third pair of leaves unfolds
5. Stem extension
6. Buds visible
7. First flower with early branching
8. Full flower and further branching
9. Late flower with most branches formed
10. Green boll; seeds are still white; and lower leaves become yellow

11. Brown boll; seeds are light brown, plump, and pliable. Branches, stem, and upper leaves are green or yellowing; middle leaves are partly dry; lower leaves have shriveled or dropped
12. Seeds ripe and brown and rattle in the bolls. Branches and upper leaves are dry but stem might still be green or yellow.

As stage one suggests, flax is an epigeal germinator. The emerging plant tissue will push the cotyledon tissue above the soil level. Should an adverse weather event such as hail or blowing dirt arise at this early stage and damage the cotyledon leaves, the flax plant may not survive.

The life cycle of the flax plant consists of a 45 to 60-day vegetative period; a 15 to 25-day flowering period; and a maturation period of 30 to 40 days. As Cullis describes, the vegetative period will see the plant produce a main stem and perhaps two or more tillers if plant density is low, soil Nitrogen levels are high, and growing conditions are favorable. The plant has a taproot (flax is a dicot plant) that can extend up to one meter into the ground. The plant stems will continue to elongate and form leaves until photoperiodicity dictates otherwise. The flowering period is characterized by the presence of purplish-blue flowers comprising five petals. Flax petals open at dawn and by early afternoon may have fallen off the plant. Flax is self-pollinating, but insects such as bees can often be seen helping the cause. After self-pollination, the ovary located at the base of the flower starts to swell. The ovary is known as a *boll* and contains the developing seeds. Boll ripening starts 20 to 25 days after flowering. A boll is divided into five compartments with each compartment containing up to two flat, oval seeds. There are two general varietal flax types: fiber flax and oilseed flax. Russia, China, and the E.U. are the main growth areas for fiber flax. Canada, India, and the US are the main growth areas for oilseed flax. Seeds from fiber flax varietals are small with 1000 seeds weighing up to 5.4 grams. Seeds from oilseed varietals are larger with 1000 seeds weighing up to 14 grams.

The straw from oil seed varietals of flax is resilient and does not readily biodegrade. Smoke billowing from piles of burning flax straw can often be seen on the Saskatchewan landscape in late autumn after harvest is complete. Plant breeding efforts are ongoing in North America to create flax varietals that will yield economic seed output while having straw that can be processed into value-added products. One such varietal is FP944, marketed as the *Klasse* varietal. Efforts are also ongoing to study the incorporation of flax fiber into industrial composites. One day, the fenders on cars could be made of flax-fiber-reinforced plastic. One day, the concrete blocks used in building foundations could be reinforced with flax fibers.

Flax seeds comprise about 41% fatty acid, 28% fiber, 20% protein, and 7% carbohydrates. The fat is typically 73% polyunsaturated, 18% monounsaturated, and 9% saturated fatty acids. A saturated fatty acid molecule can best be envisioned as a chain of carbon atoms with hydrogen atoms attached. Each carbon atom can host a carbon atom on either side of it along with two hydrogen atoms. Saturated fatty acids are regarded as potentially unhealthy. Unsaturated fats have missing hydrogen atoms in the molecular chain. The carbon atoms in the absence of the hydrogen atoms are double-bonded. An unsaturated fat is liquid at room temperature and is regarded as being healthy. Half of the fatty acids in flax oil are alpha-linoleic acid which is a digestive tract precursor to omega-3 oil fatty acid. About 18% of the fatty acids are linoleic acid which is a digestive tract precursor to omega-9 fatty acid. [5][6][7][8]

The sketches in Figures 5-1 and 5-2 illustrate the flax plant and the boll structure.

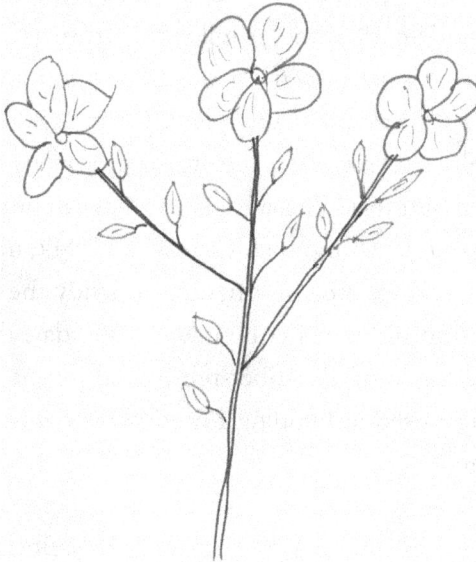

Figure 5-1
Flax Plant in Flower Stage

Figure 5-2
Flax Boll cross-section

Canola *(Brassica napa and Brassica rapa)*

- Seed Structure: Dicot
- Growth: Epigeal
- Vegetative Growth: 40-60 days (*)
- Flowering: 14-21 days (*)
- Reproductive Growth: 40-60 days (*)
  (*) Hybrid varietals exhibit shorter growth times

Rapeseed is a member of the crucifer plant family, which is the same family that includes broccoli, cauliflower, cabbage, kale, bok choy, arugula, Brussels sprouts, collards, watercress and radishes. Archeologists have uncovered evidence that rapeseed type plants were being grown in India 4000 years ago and in China 2000 years ago.

The story of canola development in Canada dates to the World War II era. Steam powered engines were integral to the war effort and rapeseed oil was used as a lubricant on cylinders and pistons. As the war dragged on, supplies of rapeseed oil from Asia and Europe were cut off.

Facing supply challenges, the Canadian government decided to mount an effort to grow rapeseed. The reasoning was that the cooler climate in parts of western Canada would be ideal for growth. A strain of rapeseed called *Brassica napus*, thought to have originated in Argentina, was selected for growing trials. Government officials also learned that a farmer from Shellbrook, Saskatchewan had brought back some seeds of the varietal *Brassica rapa* from Poland in 1936. Growing trials were successful and with some price support (6 cents per pound) from the Canadian government, western farmers started growing both varietals to meet the demand for rapeseed oil lubricant.

With the war over, the Canadian government withdrew its price support plan in 1948. This decision was further driven by the fact that diesel engines were replacing steam power and the amount of lubricant needed across industry was falling. To encourage farmers to continue growing rapeseed, the government helped identify a new market in Japan where rapeseed oil proved to be suitable for deep frying Japanese tempura style foods.

In the 1960s, focus shifted to possible human consumption of rapeseed oil. Early efforts encountered a setback when it was discovered that rapeseed oil contains *erucic acid*, a long-chain, mono-unsaturated, omega-9 fatty acid molecule not suited for human consumption. In 1968, scientists in Saskatoon, Canada analyzed a large inventory of rapeseed using gas chromatography. From this large selection, a population of low erucic acid ($< 2\%$ acid) seed was isolated. These low-acid seeds were cross-bred and by 1973 varietals of rapeseed with low erucic acid were being grown.

The rapeseed meal left over after crushing was fed to swine and poultry. However, it was discovered that the rapeseed meal contained *glucosinolates* which in the presence of water made *isothiocyanates* which interfered with iodine uptake via the animal's thyroid gland. This explained why swine and poultry species were not benefitting from eating the rapeseed meal.

By 1977, a low glucosinolate varietal had been developed and was becoming widely used. By 1980, all growers in Canada were using the low erucic acid / low glucosinolate varietals. However, the American market remained an elusive target for these modified varietals of rapeseed. Armed with ample evidence attesting to nutritional benefits, Canadian researchers were successful in having these varietals granted G.R.A.S. (Generally Regarded As Safe) status by American authorities in 1985. To add a nuance to the marketability of the product, the rapeseed name was dispensed with and replaced by the name canola. In this name, 'can' derives from Canada and 'ola' derives from the Latin expression for oil.

Today, the main varietal grown traces its lineage to *Brassica napus*. This varietal is sometimes referred to as *Argentine canola*. In excess of 25 million tonnes of canola is grown worldwide today. Canada and the USA each account for about 3.5 million tonnes of this overall figure.[9] Varietals related to Brassica napus are also referred to as *open-pollinated* varietals.

A second varietal with a shorter growing time traces its lineage to *Brassica rapa*. This varietal is sometimes referred to as *Polish canola*. In my travels as a hail adjustor in 2020 and 2021, I never heard a farm operator use the expression *Polish canola*. Rather, farm operators referred to these shorter-growth-time varietals as *hybrid canolas*.

Hybrid canola seed is developed using a combination of female and male in-bred plants. Female plants and male plants are seeded in adjacent strips in a field, with the female strips being wider than the male strips. Honey bee (*Apis mellafera*) hives are placed adjacent to the field to aid in carrying male pollen to the female plants. Sometimes leaf cutter bees (*Megachile rotundata*) are utilized. This unique strain of bees is very efficient at helping to pollinate plants. After the female plants have been pollinated, the male strips are cut down. The female plants are allowed to grow to maturity, the seeds harvested and readied for sale to growers in the next season. These resulting seeds will produce plants that bear pollen and are self-pollinating. This is due to the *fertility restorer gene* that is passed from the male parent.

This all seems like a lot of work. And it is. Hybrid canola seed is more expensive than open-pollinated canola seed. An open-pollinated canola varietal starts as seed of a particular brand. The seeds are planted, self-pollinate and go on to form seed-filled pods. The pods are harvested and readied for sale to growers in the next season.

The benefit of going to the effort to make hybrid canola seed rests with genetic improvements. By carefully selecting parental strains, seed can be developed that displays particular disease resistance, increased yield, better response to fertilizer inputs, faster growth, and perhaps even increased resistance to physical damage from hail. This type of improvement is what plant scientists refer to as plant vigor or *heterosis*.

In my travels doing hail adjusting in 2020 and 2021, I did meet farm operators who were using hybrid canola varietals. They were able to plant the seeds later in May and about 85 days later had a canola crop ready for swathing. By swathing, they were able to forgo the expense of using chemical spray to desiccate the crop prior to combining. (10)

Longer-growing-time, open-pollinated canola will produce a seedling above the soil in 7 to 20 days after planting. The key variables affecting this timeframe are air temperature, soil temperature, and soil moisture. The cotyledons which are pulled above the soil by the hypocotyl will supply the young plant with nutrition via photosynthesis. The developing roots will transport moisture and nutrition from the soil to the plant tissues. The plant will begin to develop true leaves 4 to 8 days after emergence. Leaf growth at the early stage will comprise 4 to 7 leaves in a rosette pattern.

Depending on growing conditions, the plant will go on to generate somewhere between 9 and 30 leaves on the main stem. The ideal situation is for leaves to grow rapidly. Agronomists refer to the *leaf area index* (LAI) as a measure of leaf area per unit of ground area around the plant. A LAI measure of 4 is optimal, which implies 4 square meters of leaf surface per square meter of ground surface around the plant. This level of LAI will capture 90% of the incoming solar radiation and optimize the photosynthesis process. The vegetative stage will persist for 40 to 60 days after planting. Axils at upper leaves can lead to the formation of secondary branches.

Lengthening days will trigger a shift to flowering (reproductive growth) at near 60 days after planting and small flower buds will appear on the main stem. The canola flowering stage is deemed to last from the time of first flower through to evidence of young pods on the plant structure (14 to 21 days). Flowers are yellow in color and usually comprise four petals. The flowering stage might still be evident up to 85 days after planting. Immediately ahead of the onset of flowering, nearly 60% of the plant growth has occurred.

Canola will generate more flower buds than it can ultimately support. When examining a canola plant, expect to see evidence of empty stems which are young buds that have been rejected by the plant.

The development of pods will next take 40 to 60 days depending on growing conditions. Pods will be between 1-2 inches long. The pod walls act as an additional source of photosynthesis. If a person slices open a pod, tiny, greenish, translucent seeds can be seen. Under ideal conditions, a pod can contain in excess of 30 seeds.

Weather stress events will have a notable impact on the size of the individual seeds as well as the number of seeds. As pod maturity proceeds, the coloration of the pods will shift from green to brownish. When pod moisture content is at about 30%, the crop can be swathed. Lower moisture levels are needed if straight cutting is the harvest method being used. In total, canola (Argentine varietal) will take up to 140 days from planting to being passed through a combine. (11) The sketches in Figures 5-3 and 5-4 illustrate the growth of the canola plant.

Harvested canola seeds will contain about 40% oil. The oil will contain less than 2% erucic acid and less than 30 micro moles (μmols) per gram of glucosinolate. About 60% of the oil is unsaturated oleic acid. About 10% of the oil is α-linoleic acid. Canola oil is the only plant oil that contains what is regarded as the optimal 2:1 proportion of linoleic to α-linolenic acid. The meal remaining after crushing will contain about 37% protein. Canola meal is used for animal feed and ranks third behind soybean meal and cotton meal as a source of animal feed. (12)

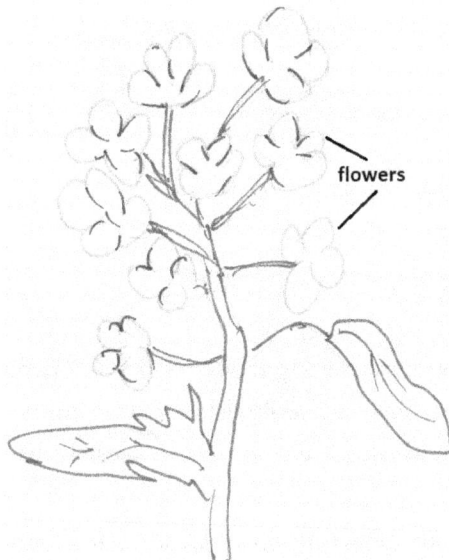

Figure 5-3
Canola Plant in Flower Stage

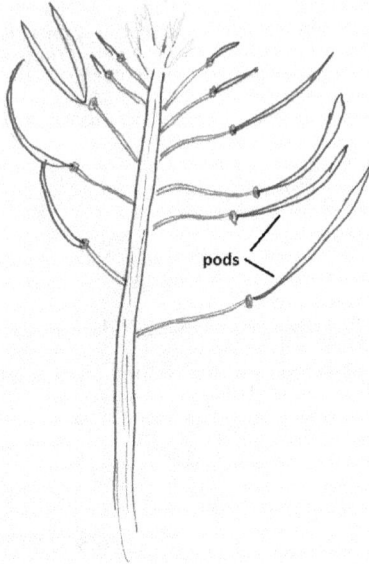

Figure 5-4
Canola Plant in Podding Stage

Mustard *(Sinapsis alba* and *Brassica juncea)*

- Seed Structure: Dicot
- Growth: Epigeal
- Vegetative Growth: 60 days
- Flowering: 10 days
- Reproductive Growth: 30 days

Sanskrit and Sumerian texts from 3000 B.C. mentioning mustard as a food item have archeologists of the opinion that mustard originated in the Indus Valley of north-west India.

The two varietals grown most often in North America are yellow mustard *(Sinapsis alba)* and brown/oriental mustard *(Brassica juncea*). Yellow mustard is otherwise known as *condiment mustard*. A commercial example is *French's Prepared Mustard* which can be found on most grocery store shelves. Coloration of seeds and orientation of pods to the stem are ways to distinguish between the two mustard varietals. Plant maturity will be reached in about 80 to 95 days, with yellow mustard being the quicker. Mustard is a crop that closely follows the growth stages of that of canola.

The largest mustard grower in the world is India. The number two grower is Canada with the bulk of Canadian mustard grown in Alberta, Saskatchewan, and Manitoba. The growing areas are usually (but not exclusively) located nearer to the US border where climatic conditions are perhaps a bit warmer and soils less prone to waterlogging from heavy rains. Brown/oriental mustard is often ground into a flour which is used in mayonnaise, salad dressing, sauces, and Dijon mustard. Varietals of condiment mustard typically contain just under 200 µmols per gram of glucosinolates and 20-25% erucic acid. If these numbers seem high relative to canola, they are. A consumer only ingests a small amount of condiment mustard at one time, which makes these elevated levels acceptable. Efforts are ongoing, however, to breed varietals with glucosinolate and erucic levels closer to that of canola as mustard oil as a cooking condiment comes into fashion. [13][14]

Legumes, Pulses, and Nitrogen Fixation

The expression *legume* denotes a podded structure. The expression *pulse crop* is used in the context of legumes. A *pulse crop* is a legume plant from which ripened seeds are harvested for human consumption. In addition to lentils, other legume plants include field peas, chickpeas, beans, soybeans, tamarind fruit, and peanuts.

A legume plant is capable of interacting in a symbiotic manner with *Rhizobia* bacteria in the soil. The bacteria senses that the plant root system has emitted *flavonoids* (polyphenol ring structures attached to a carbon atom backbone). The bacteria enter the root hairs, causing the hairs to curl. This curling action creates nodes on the root structure. The nodes act to absorb Nitrogen from the air that has penetrated to the sub-surface soil level. (Air is about 78% Nitrogen, 21% Oxygen and 1% other gasses).

The Nitrogen absorbed onto the root nodes undergoes the following reaction:

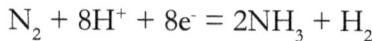

$$N_2 + 8H^+ + 8e^- = 2NH_3 + H_2$$

$$NH_3 + H^+ = NH_4^+$$

The NH_4^+ is transformed partly into amino acids, which provide nutrition to the plant. Some of the NH_4^+ is retained in the soil to provide nutrition for next year's crop.

Lentils *(Lens culinaris)*

- Seed Structure: Dicot
- Growth: Hypogeal
- Growth Period: 90-110 days

Lentils trace their origin to the Younger Dryas period and early Neolithic man's plant domestication efforts. The genus *Lens* is part of the subfamily *Faboideae* which is contained in the flowering plant family *Fabaceae*, more commonly known as the *legume* family.

Lentils are grown the world over, with annual global production in the 6.5 million tonne range. About 1.2 million tonnes of production comes from India. In Canada, annual production is usually in the 2.8 million tonne range. Output was down significantly in 2021 (to just over 2 million tonnes) as a result of dry growing conditions. The province of Saskatchewan is the epicenter of Canadian lentil production. In the US, eastern Washington and western Idaho are the focus of lentil production.

Plant researchers have identified four major genotypes of lentil: *Lens culinaris*, *Lens lamottei*, *Lenas ervoides*, and *Lens nigricans*.

Under the genus *Lens culinaris* there are 4 sub-species: *culinaris*, *orientalis*, *odemensis*, and *tomentosus*. The prevailing thought among crop scientists is that the *orientalis* subspecies is most closely related to the wild Neolithic progenitor of lentils. [15]

In 1990, USDA plant scientists Erskine, Muehlbauer, and Short published a report outlining the stages of lentil growth. These stages are now universally accepted across all lentil varietals. The motivation for this work was the *indeterminate* growth feature of lentils. The term indeterminate refers to the ability of a lentil plant to still be exhibiting some vegetative growth even as it expresses reproductive podding and seed growth.

Erskine and colleagues describe the vegetative growth stages of a lentil plant based on nodes visible on the stem. A node appears as a small, knob-like, bump on the stem. The bump occurs due to the plant expressing tissue that leads to formation of leaves and branches.

Lentils are hypogeal, which means the cotyledons of the germinating seed stay in the ground, making them less vulnerable to frost, wind erosion, or insect attack during early growth.

After germination, the epicotyl part of the stem pushes cotyledon tissue up and through the soil surface. This emergence process will take 10-20 days depending on weather and soil conditions. In Erskine's methodology, the cotyledon tissue is assigned a count value of 0.

The lentil plant will next form two nodes at or just beneath the soil surface. The leaves at these nodes are called the *scale leaves*. The sketch in Figure 5-5 illustrates some lentil growth features.

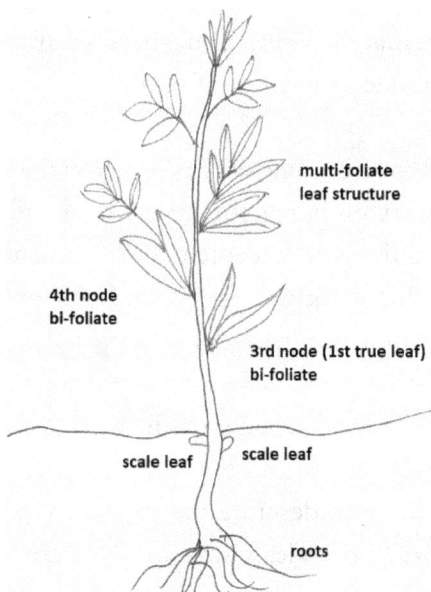

Figure 5-5
Lentil plant features

The expression of the third node gives way to the first leaf structure (the true leaf) above the soil level. Additional nodes then are formed on alternating sides of the stem structure. Beyond the third node, the next couple leaves visible on the plant stem are of a bi-foliate structure (consisting of two leaflets). Thereafter, additional leaves expressed by the plant will be all multi-foliate (more than one pair of leaflets). Under ideal growing conditions a lentil plant can add a new node every 4 or 5 days. A young leaf will start to unfurl as it matures. For counting purposes, once the two edges of the leaf no longer touch each other, a node count can be assigned.

A lentil plant can produce a single main stem and it can also produce several additional basal branches. Stem and branch expression is dependent on growing conditions. The basal branches will carry about 52% of the reproductive seed pods while the main stem structure will host about 17% of the pods. (16)

Erskine and colleagues delineate lentil reproductive stages as follows:

R1: First bloom - one open flower at any node.

R2: Full bloom - flowers open on nodes 10-13 of the basal primary branch.

R3: Early pod - pods on nodes 10-13 of the basal primary branch visible.

R4: Flat pod - pods on nodes 10-13 of the basal primary branch has reached its full length and is largely flat. Seeds fill less than half of the pod area but can be felt as a bump between the fingers.

R5: Full seed - seed in any single pod on nodes 10-13 of the basal primary branch are swollen and completely fill the pod cavity.

R6: Full pod - all the normal pods on nodes 10-13 of the basal primary branch are swollen cavities and completely fill the pod cavity.

R7: Physiological maturity - The leaves start yellowing and 50% of the pods have turned yellow.

R8: Full maturity - 90% of pods on the plant are golden-brown.

For a collection of plants in a field, the growth stage expressed by 50% of a random selection of ten plants defines the stage.

Lentils are photoperiodic plants, but do not precisely fit the short or long-day characterization. The lentil plant is unique in that its reaction to light photon energy can be disturbed by *abiotic stress* events such as temperature, strong wind, and even hail. A stress event will bring about an anatomical, physiological, reproductive, and biochemical change in the plant. Abiotic stress events can induce flowering and reproductive growth. As noted earlier, vegetative (indeterminate) growth can continue in conjunction with reproductive growth.

The number of pods per plant will vary widely, perhaps between 5 and 30 pods per plant. A pod will contain 1 or 2 seeds. The entire cycle of vegetative through reproductive growth will take 90 to 110 days depending on weather conditions.

Lentils provide everything that cereal crops don't: huge concentrations of complementary protein, vitamins A and B, iron, as well as natural nitrogen fixation to maintain soil fertility. Lentils are broadly divided into two market classes based on color: red lentil and green lentil. Each color class has three relative sizes. Red lentils are graded in the USA as extra small, small, and medium. Green lentils are graded as small, medium, and large. Large-seeded lentil types range 6 to 9 millimeters diameter; medium from 5 to 6 millimeters; small from 3 to 5 millimeters, and extra small up to 3 millimeters. [17][18][19]

Field Peas *(Pisum sativum)*

- Seed Structure: Dicot
- Growth: Hypogeal
- Growth Period: about 100 days

Field Peas is one of the earliest domesticated crops on the planet. The exact geographic origin has not been identified but archeologists have found trace remnants of peas across the Mediterranean region in general, and in the Fertile Crescent region (modern day Iran and Iraq) in particular. Archeological evidence shows that early man was consuming peas at 10,000 B.C. Evidence suggests that from the Fertile Crescent, peas spread to Russia and thence to Europe and also to India and China.

Global cultivation of peas ranges between 1.6 to 2.2 million planted hectares (4–5.4 million acres). Annual production is between 12–17 million tonnes per year.

Peas are a rich source of protein (23–25%), essential amino acids, complex carbohydrates, as well as iron, calcium, and potassium. Peas are naturally low in sodium and fat. Peas are used in soups, breakfast cereals, processed meat, health foods, pasta, and purees; they are also processed into pea flour, starch, and protein.

Three species of peas are grown globally today:

Pisum sativum extends from Iran and Turkmenistan through Asia, northern Africa, and southern Europe.

Pisum fulvum is found in Jordan, Syria, Lebanon, and Israel.

Pisum abyssinicum is found from Yemen and Ethiopia.

In North America, it is the *Pisum sativum* species that is grown under the industry name of Field Pea. End uses are for both human consumption and animal feed.

Pisum sativum is a legume plant characterized by weak stems that prevent it from growing much beyond 60 centimeters (2 feet) in height. The leaves are comprised of leaflets and tendrils. Flowers are white, pink, or purple. Pods carry seeds (peas) that are nearly spherical in shape, and white, gray, green, or brown in color. *Pisum sativum* is a hypogeal plant with the cotyledons remaining at or below the soil surface. [20]

Plant breeding efforts are ongoing in Canada and the US to develop pea varietals with better yield, improved resistance to blight, improved color retention, and improved resistance to lodging (crop laying down). Semi-leafless pea varietals that carry the *afila* gene will produce tendrils instead of leaflets. The tendrils help to prevent lodging. [21][22]

Plant scientist Kristen MacMillan of the University of Manitoba describes the pea growth process as follows, where the prefix V denotes vegetative growth and the prefix R denotes reproductive growth: [22]

V0: the epicotyl structure has pushed up through the soil

V1, V2: two small scale leaves appear on the stem

V3: 17-21 days after planting

V4-V5: 24 to 34 days after planting

V8: 30 to 46 days after planting

R2: (first flowers): 49-55 days after planting

R3: (first pods): 52 to 64 days after planting

R4: (podded): 61 to 70 days after planting

R5: 70 to 77 days after planting

R6: 74 to 84 days after planting

Harvest: 90 to 97 days after planting

The sketch in Figure 5-6 illustrates pea plant features.

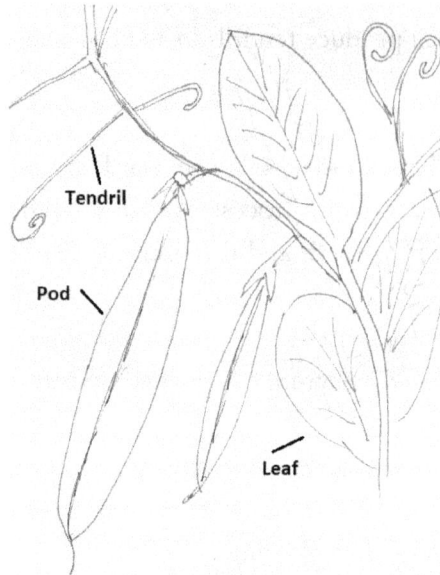

Figure 5-6
Pea Plant

Chick Pea *(Cicer arietinum)*

- Seed Structure: Dicot
- Growth: Hypogeal
- Growth Period: about 100-130 days

The chick pea (*Cicer arietinum*) is a legume plant. Chick peas are sometimes referred to as garbanzo beans. Archeologists have uncovered chick pea remains that date to 9500 years old in southeast Turkey. From this origin point, evidence suggests chick peas reached Bulgaria and Greece via the Black Sea. Thereafter, chick peas found their way to India and also down the Nile River into Africa. In 1492, chick peas found their way to the New World with the explorer Columbus.

Chick pea seeds are high in protein. They are a key ingredient in hummus and chana masala, and can be ground into flour to make falafel. Chick peas are also used in salads, soups, stews, and curry. India is responsible for about 68% of the 13 million tonne global chickpea production. Canada is a small player, producing only about 250,000 tonnes annually.

Chick pea (*Cicer arietinum*) is divided into two cultivar groups: Desi and Kabuli. White flowers, a thin seed coat, a large, cream-colored seed (200–680 mg) with a smooth surface and lack of anthocyanin pigmentation define the Kabuli chickpea. Pink-purple flowers, a thick seed coat, a small, angular, dark seed (100–200 mg) with a rough surface and anthocyanin pigmentation define the Desi chickpea.

Chick pea plants are sensitive to the environment and will neither tolerate hot weather, drought, or waterlogged soils. The time of flowering onset depends on photoperiodicity (day length) and temperature.

Chick peas comprise about 60% carbohydrates, 19% protein, and 17% fiber. Recent research has revealed that chick peas contain bioactive compounds such as biochanin, genistein, trifolirhizin, calycosin, formononetin, sissotrin, and ononin. These unusual sounding compounds have potential to confer antifungal, antioxidant, estrogenic, insecticidal, and antimicrobial benefits to the person eating the peas.

The growth characteristics are similar to that of the Field Pea with a new node produced on the main stem every 3 or 4 days under suitable growing conditions. Main stem nodes then express up to 7 or 8 primary branches which in turn lead to numerous secondary branches and leaves. Like the lentil, the chick pea will display indeterminate growth. Flowering will begin around the 13th or 14th node, which equates to 55 to 75 days after seeding. The plant is self-pollinating and each pod contains one or two peas. Harvest maturity occurs 100 to 130 days after seeding, depending on location and growing conditions. The sketch in Figure 5-7 illustrates a chick pea plant.

The chick pea plant is capable of secreting malic acid which is a deterrent to insects. However, the Ascochyta blight fungus remains a threat to chick peas and can lead to severe crop failure.

Farm operators often will exercise a 3-to-5-year rotation pattern when growing chick peas. [23][24][25]

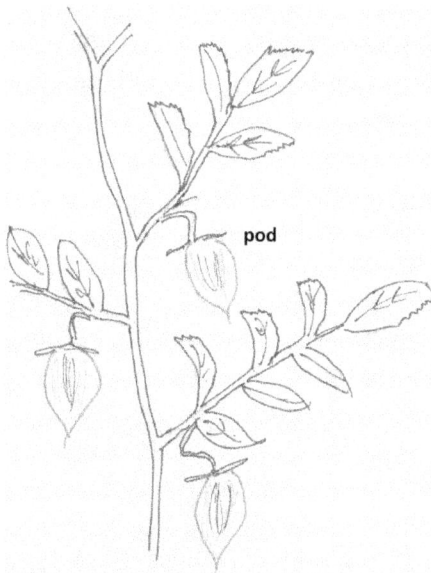

Figure 5-7
Chick pea plant

Corn *(Zea mays)*

- Seed Structure: Monocot
- Growth: Hypogeal
- Growth Period: about 120-130 days

Columbus noted the presence of maize (corn) in 1492 as he set foot on the shores of what is today Cuba. What Columbus did not realize was that maize was already growing throughout North America and South America.

Historians and archeologists generally agree that maize (corn) traces its lineage to nearly 11,000 years ago. Scientists believe that at that time, there was a thin strip of land connecting what is today northern China to what is today Alaska. Today, this strip of land has been replaced by the Bering Sea. Hunter/gatherer early man made the journey across this land bridge, bringing with him seeds of annual and perennial grasses. Once planted, these seeds soon took root and began to proliferate.[26]

Genetics researchers have determined that the wild progenitor of maize was a seed-bearing grass called *teosinte*. As the climate continued to warm and evolve, it appears likely that teosinte evolved as well. Over time and through human plant breeding intervention, maize has evolved into what we now broadly call *corn*. Several hundred varieties exist around the world ranging markedly in height and kernel coloration. [27]

A 2006 article from the Japan Forage Crop Research Institute describes how corn is different from other cereal grains in that the female (ear) and male (tassel) are located somewhat apart from one another on the plant. The male tassel is situated above the female ear. The reproductive growth (cob) will contain 300 to 1000 kernels in 12 to 16 rows. The seed coat on a kernel is near 87% fiber. The endosperm of a kernel is near 87% starch and 8% protein.

Corn is a significant source of food nutrition for humans and animals. Moreover, corn is a source of starch, oil, and protein.

Starch (molecular chain of glucose molecules) is amenable to being fermented by yeast. This is why corn is heavily used in the beverage alcohol industry. In my distillation consulting efforts, I always recommend corn as part of an alcohol creation recipe. The entire Kentucky bourbon industry revolves around corn with many recognized bourbon brands containing 70% or more corn in their recipes. The next time you are enjoying a shot of *Jack Daniels* whisky, you are drinking distillate made from a recipe containing 80% corn and 20% other cereal grains. In Canada, *Crown Royal* whisky and *40 Creek* whisky derive their unique flavor profile partly from the corn in their respective recipes. Corn even extends into the vodka industry. Popular vodka brand *Tito's* is based on distillate made from corn. Many vodka products are nothing more than brand names. The distillate used to create these products has been sourced in bulk from an industrial scale ethanol factory. By Canadian and American legal definitions, vodka is nothing more than ethanol diluted down with water to a drinkable strength of around 40% alcohol by volume. Through clever marketing programs, all too often consumers are led to believe that the vodka they have purchased has been hand crafted by a dedicated artisan master distiller.

Corn is also used as feedstock for creating ethanol additives for the fuel industry. The next time you are filling your vehicle with gas, you might notice a small sticker on the pump that advises you the gasoline contains up to 10% ethanol.

Corn is also used in making sweeteners. The next time you are in the grocery store, take a moment to look at the ingredients in food items. You will notice the near ubiquitous presence of high fructose corn syrup. This is starch derived from corn that has been treated with enzymes. To the human taste buds, fructose tastes sweeter than the sucrose sugar derived from sugar cane or sugar beet plants. This gives food producers the luxury of controlling production costs (and raising profit margins) by using smaller quantities of high fructose corn syrup in products. The problem is, the human digestive system does not readily assimilate fructose and will store it as fat in the body. In recent years there has been a dietary pushback against fructose with people seeking out products made with the diet conscious substance *erythritol*. To make erythritol, corn is processed with enzymes to release the starch content. The starch is then fermented with a special yeast strain such as *Yarrowia lipolytica*. The net result is a crystalline substance called erythritol which has far fewer carbohydrates than sugar, yet tastes every bit as sweet. (28)

Stages of corn growth are well documented by seed providers such as Bayer Crop Science and by institutions such as Iowa State University:

After seeding, it takes 5 to 14 days before coleoptile plant matter is visible and emerged above the soil. This emergence is denoted as the **VE stage**.

The counting of growth stages revolves around what is called the *collar method*. Just as the collar of your shirt wraps part way around your neck, a leaf will wrap part way around the corn stalk structure. It is only when a leaf exhibits this wrapping feature that it is assigned a stage count (V1 to Vn).

V1: one leaf shows collaring.

V2: two leaves show collaring. The plant is 50-100 mm (2-4 inches) tall and is still using energy from the seed endosperm via the scutellum (cotyledon) membrane. The seedling roots are elongating.

V3: this stage arrives about two to four weeks after VE. The plant switches to photosynthesis for nutrition.

V4: leaf and ear shoot tissue material are starting to form.

V5: the plant is 20-30 cm (8-12 inches) above the ground, but the primary growth point is still beneath the soil as the root system is expanding.

V6: occurs four to six weeks after VE. The plant growth point is now above the soil.

V7: stem elongation becomes evident.

V8: the plant will be 60 cm (24 inches) tall.

V9: the development of the tassel will be evident.

VT: denotes terminal growth. This stage occurs about 9-10 weeks after VE and is marked by the emergence of silks.

Following these vegetative growth stages, pollen shedding will next carry on for up to two weeks. The timeframe on either side of silking is critical to corn growth. This timeframe is when the plant needs maximum moisture. This is also the time when the plant is most susceptible to damage from storms, including hail storms.

The reproductive stages (**R1** to **R6**) begin at **VT**. Progress is marked by kernel development.

R1: a field is at stage R1 when 50% of the plants are exhibiting silking. Maturity of the plant will be complete about 60 days after R1.

R2: will be complete after about 12 days. The kernels appear as small, water-filled, blisters.

R3: will be complete about 20 days after R1. The kernels will appear yellow and will be filled with a milky, white substance.

R4: will be complete about 26 days after R1. The substance in the kernels will have a dough-like composition.

R5: will be complete about 38 days after R1. The kernels will exhibit a dent formation. If the crop is being grown for silage, harvesting will begin in this timeframe.

R6: denotes maturity as is sometimes called the *black layer stage*. The tips of the kernels will take on a darker coloration. This stage will be complete about 60 days after R1.

Adding up these typical intervals shows that the overall growth of a corn plant will be 120-130 days. This timeframe will be varietal and growing-condition dependent. [29][30][31]

Wheat, Barley, Rye, Oats and other Cereals

- Seed Structure: Monocot
- Growth: Hypogeal
- Growth Period: about 120 days

Cereal grains trace their origins to about 12,000 years ago in the Fertile Crescent, a region today that encompasses Turkey, Iran and Iraq. The expression *cereal* derives from Ceres, the goddess of crops and agriculture in ancient Roman society. [32]

One of the earliest varietals of cereal grain was *emmer*. The modern day, cultivated version of emmer traces its lineage to a wild, natural cross breeding between grass species *Triticum Urartu* and *Aegilops speltoides*. Modern day durum wheat is closely related to wild emmer.

Another early variant of cereal was einkorn (*Triticum boeticum*) which has single grain kernel on each spikelet. Modern day farro wheat grown in Italy traces its lineage to einkorn.

Modern day bread wheat is a complex chromosomal mix derived from *Triticum urartu, Aegilops speltoides*, and *Aegilops tauchii.* [33]

The evolutionary path of rye grain is not entirely clear. It is thought that to have originated as a weed. Early man collected its ripe seeds and planted them. The domesticated version of rye is *Secale cereale* and is relatively tolerant to frost, drought, and marginal soil fertility.

In my distillation consulting, I always make an effort to clarify the rye situation. Consumers have a habit of describing the amber-colored liquid in whisky bottles as rye. Nothing could be further from reality. While some whisky products might have been made from a grain mash recipe containing rye, the amount of rye in the recipe is small. There are small-batch craft distillers who do make whisky from recipes containing exclusively rye. The taste of a true rye whisky is far different from a typical commercial rye whisky product. In Canada, there are no legal guidelines for rye whisky. A distiller in western Canada (I shall refrain from mentioning who) incorporates about 10% rye and 90% wheat into a recipe and still manages to call its product rye whisky. In the US, the guidelines are more explicit. A US distiller must use a minimum of 51% rye grain in a recipe in order to be able to call the resulting distillate a rye whisky.

The evolution of oats traces back to a seed-bearing, wild grass, *Avena sterillis*, which evolved into a couple hundred different sub-species. Oats are about 66% carbohydrates, 11% dietary fiber, 4% beta-glucans, 7% fat, and 17% protein. Of all the cereal grains, the oat was the last to be purposely grown for human consumption. This effort started in the 1500s as Scandinavian countries started growing oats. Credit goes to the Scots for eventually turning oats into a daily food product. The English, however, curled their noses at the mention of oats. Englishman Samuel Johnson is reported to have written in 1755 that oats were "eaten by people in Scotland but fit only for horses in England." A Scottish Lord is said to have retorted, "That's why England has such good horses, and Scotland has such fine men!" (34)

Cereal grains are hypogeal. The cotyledon structure remains under the soil as growth gets underway. Growth stages of cereal grains is typically expressed by the Zadoks scale, developed by Dutch cereal crop scientist Jan Zadoks in the early 1970s. The scale extends from 00 through 99.(35)

Zadoks 00-09: From seeding to emergence takes up to about 9 days. Coleoptile tissues will be evident just above soil level. Wheat germination requires the seed to be at 35 to 45 percent moisture by weight. The optimal soil temperature for germination is 12° to 25°C.

Zadoks 10-19: From 10-19 days, the plant will exhibit seedling growth. Crown nodes (nodes that do not elongate) will develop on the plant just under the soil surface.

Zadoks 20-29: From day 20-29, the plant will express tillers (lateral shoots), if temperature and moisture conditions allow. Tillers form at the plant crown nodes. Tillering usually starts when the plant has expressed about three true leaves. The base of each tiller is protected by tissue called the *prophyll*. Tillers are dependent initially on the main plant stem for nutrition. Once a tiller has expressed about three leaves of its own, it will develop its own root system and begin to grow.

Zadoks 30-39: From day 30-39, the plant expresses stem elongation. This is sometimes referred to as *jointing*. The maximum possible number of kernels per head is determined by the plant at this time based on soil and growing conditions. The plant will allocate nutrients to the main stem and tillers with at least three leaves. Once the plant has started jointing, no more tillers will form. If the plant has been damaged during stem elongation as a result of hail, frost, or insects, the main stem and tiller will die. The plant can compensate for this loss by expressing new growth from the base of the plant. Towards the end of jointing, flag leaf tissue will be visible as it forms a collar around the stem. The flag leaf is the final leaf formation in the growth process. The flag leaf with its surface area and location on the plant is responsible for generating the photosynthesis nutrition needed for the head to grow and develop filled kernels. The flag leaf must be protected from diseases, damage (hail), and insects if the grain head is to develop.

Zadoks 40-49: At this point, the developing head is protected by the flag leaf. As the head grows, the flag leaf takes on a swollen appearance and resembles a boot. This stage is commonly called the *boot stage* of growth.

Zadoks 50-59: This interval begins when the tip of the head can be seen emerging from the flag leaf sheath. This interval is complete when the head fully emerges from the flag leaf sheath.

Zadoks 60-70: This interval represents the transition from vegetative growth to reproductive growth. In some cereal grains, flowering and pollination might actually commence in the waning days of the prior stage. Pollination takes only a handful of days to complete. The sketch in Figure 5-8 illustrates the pollen-producing anthers connected to the filaments. The pollen-receiving stigma is connected to the ovary structure. Protection is provided by the lemma and palea tissues.

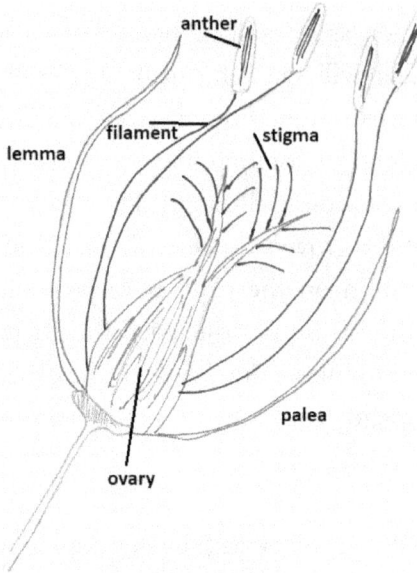

Figure 5-8
Cereal grain reproductive organs

The structural member upon which grain seed (caryopsis) occurs is called the *rachis*. Spikelets will develop on the rachis, with each spikelet having up to three florets. Spikelets alternate on either side of the rachis structure. Each spikelet will have a series of between two and six florets within. Cereal grains (durum, rye, 2-row barley) having 4 sets of chromasomes (tetraploid) will express two or three florets. Cereal grains (bread wheat, oat, 6-row barley) having 6 sets of chromasomes (hexaploid) will express six florets.

The sketch in Figure 5-9 illustrates a seed (caryopsis) that is growing with protection from the lemma, palea, and glumes.

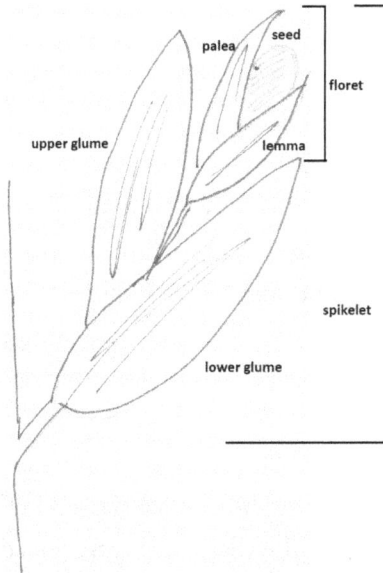

Figure 5-9
Cereal grain spikelet and floret

Zadoks 71-77: This interval is referred to as the *milk stage*. The milky fluid evident in a kernel will eventually translate into hard kernel material. Stress on the plant at this stage will negatively affect the size and weight of kernels.

Zadoks 78-90: During this interval, the milky substance will start to congeal into what is termed *soft dough*. It will then firm up into what is called *hard dough*.

Zadoks 91-99: This final stage is when the moisture level of the kernels decreases from about 35% down to about 15%. After this level of moisture is attained, the crop is said to be *combine ready*. (36)

Soybean *(Glycine soja)*

- Seed Structure: Dicot
- Growth: Epigeal
- Growth Period: about 90-150 days

The Soybean is a legume that traces its lineage to south-east Asia. The ancient ancestor *Glycine Willd* has two sub genera, *Glycine* and *Soja*. It is *Glycine Soja subsp. Glycine max* that is the modern day cultivated Soybean.

Soybean cultivation is thought to have started in China about 3000 years ago. In the 1600s, the plant was introduced to Europe. The soybean plant was brought to the British Colony of Georgia in the mid-1700s. In the mid-1900s, the plant was introduced to South America. Today, the US, Brazil, and Argentina account for 80% of world soybean production. US soybean growers produce over 4 billion bushels each year with the state of Illinois holding the distinction as the leading producer. Canada pales by comparison with just over 200 million bushels grown annually. Ontario and Quebec are responsible for the bulk of this output. The province of Manitoba accounts for 14% of this output while Saskatchewan accounts for less than 1%.

Cultivated soybeans exhibit wide variability in terms of seed shape, size, color, plant morphology and maturity. Soybeans comprise 40% protein, 20% oil, 35% carbohydrate, and 5% ash. Soybeans are the world's primary source of vegetable oil. The remaining meal left after crushing is incorporated into animal feed. The complete list of other soybean applications is too numerous to mention here. Suffice it to say that soy is an economic filler. I have a habit of looking at ingredient labels on food items as I shop. I have seen soy as a component in breads, sauces, and even processed cheese slices. Those little coffee creamers you are given at the local coffee shop contain soybean oil that resembles milk. Soaps, shampoos, paints, cosmetics, building materials are all part of the long list of soy uses. And, of course one cannot overlook biodiesel made from soybean oil.

The oil content of a soybean comprises five fatty acids: palmitic, stearic, oleic, linoleic, and linolenic. The first two in this list are saturated fatty acids and represent 15% of the oil content. The latter three are unsaturated fatty acids and account for 85% of the oil content. Plant breeding efforts are these days are focused on increasing yield but also on lowering the saturated fatty acid content and raising the unsaturated content. In addition to the oil content, soybeans contain B vitamin, folic acid, and isoflavones which are though to impede the progress of cancer, heart disease, and osteoporosis (bone density reduction).

In 1996 Monsanto created the *Roundup Ready (RR)* soybean by genetically engineering a gene segment from the soil bacteria *Agrobacterium tumefaciens* into the soybean genome. This advancement allows a farm operator to top dress the crop to kill weeds while not harming the soybean plants. (37)(38)(39)

Soybeans are broadly classed into *Indeterminate* and *Determinate* varietals. Indeterminate varietals can generally be found in northern soybean growing areas of North America. These plants are characterized by their ability to exhibit vegetative growth while also exhibiting reproductive growth. Determinate varietals typically are found in southern growing regions of the USA. These varietals will complete vegetative growth prior to exhibiting reproductive growth.

The University of Iowa, in its publication *Soybean Growth and Development*, explains the stages of soybean growth. The University of Minnesota also has excellent information on its learning extension website. The Soybean growth stages are as follows:

VE: The cotyledons have been pulled above the soil surface by the elongating hypocotyl tissue. The cotyledons supply nutrition for more growth and they also begin to capture photosynthetic energy. This stage takes 5 to 15 days from planting.

VC: The first set of leaves above the cotyledon leaves are unifoliate (comprised of one leaflet). At this stage, unifoliate leaves are evident. Subsequent sets of leaves are trifoliate (comprised of 3 leaflets). This stage will be complete in about 10 days.

V1: The first trifoliate leaf appears.

V1 to Vn: Subsequent stages are numbered as successive trifoliate leaves develop. From one stage to the next will take about 5-8 days.

The transition to the reproductive phase is denoted by the expression of at least one flower on the plants. The reproductive stages comprise R1 through R8.

R1: A plant will express at minimum one flower.

R2: Plants will express full flower. This stage will be completed after about 5-15 days from the start of R1.

R3: The start of podding will be evident. In about 5-15 days from stage R2, plants will all express podding.

R4: Full pods will be evident. This stage will take about up to about 26 days from the end of R3.

R5: Seeds visible inside pods. Seeds will be about $1/8^{th}$ inch long. This stage will take about 10-20 days from the end of R4.

R6: Podding complete. This stage will take about 9-30 days from the end of R5.

R7: Pods start to change color as they mature. This stage will take perhaps 7 to 18 days from the end of R6.

R8: 95% of pods have changed color and matured.

Obviously, these intervals have a wide distribution. Moisture, temperature, and general growing conditions will determine the rate of growth.

The sketch in Figure 5-10 illustrates the soybean plant structure in the V1 vegetative stage. [37][38][39][40][41]

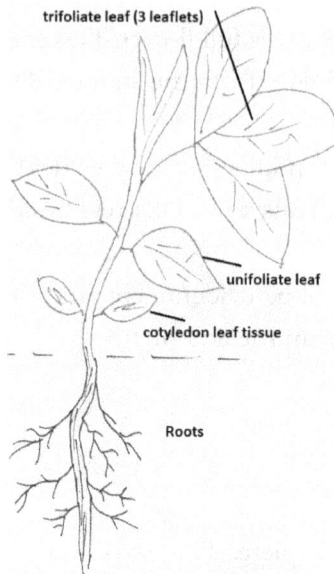

Figure 5-10
Soybean plant structure

Canary Seed *(Phalaris canariensis)*

- Seed Structure: Monocot
- Growth: Hypogeal
- Growth Period: about 105 days

Canary seed, as the name implies, is a cereal grain grown primarily for bird consumption. The plant traces its origins to the Mediterranean region. Prior to the late 1990s, canary seed was only suited for bird consumption owing to the presence of hair-like fibers called *trichomes* which are attached to the seed hulls. These fibers were deemed potentially carcinogenic to humans in a 1980 study in Iran. The trichomes are not easily separated during seed processing and cleaning. Moreover, cereal crops grown adjacent to fields of canary seed can adsorb wind-blown bits of trichome material, rendering the cereal crop harmful to human consumption.

Figure 5-11 shows mature canary seed heads. The actual seeds are only visible upon close visual examination.

Figure 5-11
Mature Canary Seed Heads

In 1997, Dr. Pierre Hucl, a plant breeder from the University of Saskatchewan's Crop Development Center (CDC), developed a hairless canary seed (*CDC Maria*) in which the trichomes were completely absent from the hull and glumes. The new hairless, or *glabrous*, species did not invoke an itching sensation on the skin when handled. In 2015, the US Government granted *CDC Maria* G.R.A.S. (Generally Regarded As Safe) status. In 2016, Health Canada granted CDC Maria novel food status.

Hairy varieties grown include *Keet*, *Cantate* and *Elias*. Hairless varietals include *CDC Maria* as well as *CDC Togo*, *Bastia*, *Calvi*, *Cibo* and *CDC Lumio*.

The province of Saskatchewan remains a significant source of canary seed with over 400,000 acres seeded each year. This acreage provides about 65% of global canary seed and accounts for over 90 percent of Canada's canary seed. The area around Kindersley, Saskatchewan is particularly noted for its canary seed production.

A 2010 study by University of Guelph food scientist Dr. Abdel-Aal showed that the endosperm (white flour) material from *CDC Maria* seeds comprised about 57% starch, and 22% protein. By comparison, a sample of Red Spring Wheat (*Katepwa* cultivar) had 74% starch and 17% protein. This information prompted a series of follow-up studies. In a 2021 study, Abdel-Aal showed that *CDC Maria* contains about 7% oil and 15-20% fiber. Moreover, only a small percentage of the seed proteins are glutelin. This protein is the one of the proteins (the other being gliadin) that can cause severe allergic reactions to people who are gluten sensitive. Although canary seed is technically deemed by Health Canada to be a gluten-free product, Health Canada does advise that people with a severe sensitivity to gluten should approach canary seed carefully.

Further to the 2021 study, *CDC Maria* was shown to comprise 53.8% linoleic acid, 24.2% oleic acid, 11.5% palmitic acid, and 2.8% linolenic acid. Linolenic and oleic acids are unsaturated fatty acids, making them desirous from a dietary health perspective. The unsaturated : saturated ratio of the *CDC Maria* fatty acids was 6.4 : 1. Olive oil has a smaller ratio of 5.8 : 1. Canola oil is superior at 12.8 : 1. The potential to incorporate small bits of *CDC Maria* canary seed into traditional baking recipes is thus significant. In addition, *CDC Maria* was shown to contain phenolic acids, lutein, zeaxanthin, and beta-carotene, all of which are beneficial to overall health as they boost antioxidant activity, improve immune response, suppress free radicals and lower the risk of cardiovascular disease.[42][43][44]

As an aside, in 2019 I was offered a small sample of *CDC Maria* canary seed courtesy of the Saskatchewan Canary Seed Development Commission. In an amber ale beer recipe created from 5.5 kgs of grain, I incorporated 10% by weight of canary seed into the beer mash recipe. All other brewing parameters were unchanged. The resulting beer was pleasant to drink. In my opinion, canary seed is something the craft beer industry should consider. Canary seed is not just for the birds anymore. As of August 2021, canary seed was designated an official grain by the Canadian Grain Commission. Growers might be able to access to some tax credits from the Canada Revenue Agency. [45]

Canary seed will mature in just over 100 days. The seed prefers soils that retain moisture, as the crop is less drought tolerant than wheat. Canary seed is susceptible to chemical residue from the previous year's crop as well as to types and amounts of fertilizer applied during seeding. Canary seed plant stems are prone to bending and arcing under the influence of wind and heavy rain. This feature is often mistaken for hail damage given that hail storms can be accompanied by high wind and heavy rain. Canary seed heads are tightly packed structures as Figure 5-11 shows, making it difficult for hail to do severe damage to a canary seed crop in the first place. While hail adjusting in 2021 in the Kindersley, Saskatchewan area

I found very little hail damage on canary seed crops I examined. This despite other crop types in nearby fields displaying significant hail damage.

Faba Bean *(Vicia faba)*

- Seed Structure: Dicot
- Growth: Hypogeal
- Growth Period: about 110-130 days

One of the more unique legume crops is the Faba bean. This bean crop traces its ancient origins to what is modern day southern Israel. Of all the bean type crops grown, Faba has a high protein content (>25% protein, 10% fats, >40% carbohydrates). A popular end use of Faba bean is for plant-based meat substitutes such those produced by US companies Ingredion (NYSE:INGR) and Beyond Meat (NYSE:BYND).

At about 30 days after seeding, a scale leaf will emerge at the first node beneath the soil level. A second node will appear just above the soil level. The third and fourth nodes will comprise a pair of leaflets. From node five and upwards, the leaves will comprise two or three pairs of leaflets. The mature bean plant will exhibit firm, hollow, stems with a total height of 0.5 to 1.8 meters tall.

Flowering occurs in 45 to 65 days after planting. Flowers can be pure white, purple or pink and each cluster may produce one to six pods.

The sketch in Figure 5-12 illustrates a Faba bean plant with two pods attached.

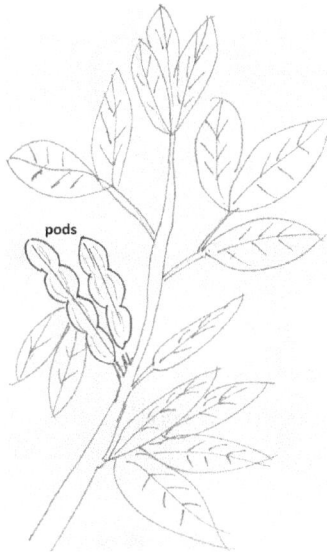

Figure 5-12
Faba bean Plant

Pods are long and green, growing up to 10 cm (4 inches) long, 1 to 2 cm (0.4 to 0.8 inches) wide, and contain two to eight round-shaped seeds per pod. The texture of a pod resembles canvas material. Pods turn brown or black as they mature and seeds turn tan to brown to grey. Faba beans prefer moisture, so in a hot, dry summer the crop is likely to suffer. [46][47]

Although grown for human consumption, Faba beans do contain the tannin compounds *vicine* (V) and *convicine* (VC). These compounds can cause life threatening oxidative stress to red blood cells. Globally, there are over 400 million people who are deficient in the G6PD bodily enzyme. This enzyme, if lacking in the body, will permit vicine and convicine to oxidatively stress red blood cells. The result will be anemic shock (hemolysis). Cooking the Faba beans or thoroughly drying them before use alters the chemical structure of the tannins and eliminates the potential for any toxic hemolysis reactions. [48]

Buckwheat *(Fagopyrum sagittatum)*

- Seed Structure: Dicot
- Growth: Epigeal
- Emergence: 6-10 days
- Flowering: 30-45 days
- Seeds Forming: 40-60 days
- Seeds 80% ripe: 75-90 days

Despite what the name suggests, buckwheat is not a wheat and not a cereal grain. It is classed as a 'pseudo-cereal'. Buckwheat is grown primarily in North Dakota, New York, and Pennsylvania. In Canada, the provinces of Manitoba, Ontario, and Quebec grow buckwheat. There was a time when the primary end market for buckwheat was Japan where it was used in making soba noodles. Demand in North America is now on the increase as consumers seek out gluten-free, nutritious, ancient grains.

Archeologists suggest buckwheat was grown in China before 1000 A.D. It was introduced in Europe in the 15th century and the U.S. in the 17th century. The name buckwheat originated from the Anglo-Saxon words boc (beech) and whoet (wheat). The three-sided, angular seed resembles a small beechnut. Buckwheat grows best in a cool, moist climate under a wide range of soil conditions. It is sensitive to spring and fall frost, high temperatures, drying winds and drought.

Stress factors such as wind, heavy rainfall, and excessive soil

Nitrogen can reduce crop yields. Buckwheat grows well in fields with lower fertility, and particularly on low-Nitrogen and low-Phosphorus soils.[49][50]

As an aside, in my recipe development work for craft distilleries, I have discovered that buckwheat added to a mash recipe of other cereal grains makes for a delicious tasting whisky distillate. In response to demand for more gluten-free alcohol products, in 2016 the corporate parent (Latvian-based SPI Group) behind *Stolychnya Vodka* introduced a gluten-free vodka version made from corn, oats and buckwheat. As part of my recipe development services offered to craft distillers, I have made blends of oat, corn, buckwheat distillate to demonstrate the impressive flavor profile.

Buckwheat is deemed to be a pseudo cereal because of its distinct differences from a true cereal grain. A cereal grain is a monocot, whereas buckwheat is a dicot. In a cereal, the endosperm cell walls comprise three layers. In buckwheat the endosperm cell walls are thin. In a cereal, the endosperm cells contain small and large starch globules. In buckwheat, the globules are all small-sized.

In a cereal, such as wheat, the endosperm storage proteins are mainly prolamins. These proteins cause reactions in coeliac sufferers who are sensitive to prolamins. Buckwheat proteins comprise 64% globulins, 12% albumins, 8% glutelins and only 3% prolamins. White bread has a glycemic index of 100. Boiled buckwheat groats (dehulled kernels) have a glycemic index of only 61. Buckwheat also contains all nine of the essential amino acids (the aminos that the human body cannot generate and thus must be sourced from the food we eat). Buckwheat, without doubt, is a beneficial product for humans to consume. Furthermore, buckwheat further contains 37% oleic acids and 39% linoleic fatty acids. These are the beneficial unsaturated fatty acids. The ratio of unsaturated to saturated is 4:1. While less than canary seed and less than olive oil, this ratio suggests that buckwheat has significant health benefits. [51]

The USDA defines a growth stage on buckwheat as being the expression of a node and a leaf on the growing stalk. Cotyledon nodes will appear above the soil in about 10 days after planting. Seven days later, Node 1 (stage N1) will appear. N2 will follow in seven more days and N3 in seven more. At stage N4 flowering starts and continues through to stage N7. The buckwheat plant is indeterminate in its growth in that a plant can express vegetative growth while also flowering. Seed begins to set at stage N8. Seed growth will continue for stages N9 and onwards. The crop is harvest ready when 70% of the seeds have turned dark brown. [52]

Sunflower *(Helianthus annuus)*

- Seed Structure: Dicot
- Growth: Epigeal
- Emergence: about 11 days after planting
- Pollination Completed: about 80 days after planting
- Maturity: about an additional 30 days
- Growth Period: about 110-130 days

Archeologists have concluded that the epicenter for sunflower origination is northern Mexico. It is estimated that the early residents of this area domesticated the sunflower around 3000 B.C. To these early peoples, the sunflower was revered for its medicinal properties. It was used as a diuretic. It was also used as an expectorant to aid people suffering from lung and throat infections. Sunflower also had practical properties. The oil extracted from the seeds was mixed with the yellow petals to make colored face painting dyes for ceremonial celebration. The plant stalks were used to make ropes and other building materials.

Settlers arriving in the New World in 1586 observed Indigenous people growing sunflowers near what is today Roanoke, Virginia. Archeologists are of the opinion that sunflower seeds were carried to eastern North America by Indigenous people from Mexico seeking trading opportunities with other Indigenous groups.

In the late 1600s, the sunflower plant was taken to Russia where the climate proved amenable to good growth. In 1716, a patent was granted in England for the extraction of oil from imported Russian sunflower seeds. By 1830, sunflower oil was being made commercially in Russia. In the late 1700s, Ukrainian farmers of Jewish origin emigrating to America took sunflower seeds with them. In the late 1800s, Ukrainian farmers of Jewish origin emigrating to Argentina also took sunflower seeds with them.

From the 1930s through to the 1960s, the focus of plant cross-breeding efforts was on increasing the oil content of the seeds. These plant breeding effort were led largely by Soviet scientists. By 1965, the oil yield on sunflower had reached 550 grams/kg. Before the plant breeding efforts had started the average yield was 330 grams/kg.

As of 2002, Argentina was responsible for 18% of global sunflower output. Since then, output has dropped off as farm operators have shifted to growing soybeans, a more financially lucrative crop. Today, Russia and Ukraine account for 50% of the global sunflower output. The US ranks 9th on the list of producing countries. About 75% of sunflowers grown in the US are grown in North Dakota, South Dakota, and Minnesota. Total US sunflower acreage is near 3 million acres. The vast majority (90%) of the sunflower crop is grown to obtain sunflower oil. The balance is for edible sunflower seeds. Edible sunflower varieties are recognizable by their larger seeds which are typically striped or whitish in color. Oilseed varietals are almost always black in color. In Canada, the province of Manitoba is the focal point of sunflower growth with 36,000 acres grown in 2020. Sunflower yields can be 2000 pounds per acre in a good season. The oil content of a sunflower seed is about 40%. After oil extraction, the remaining meal is marketed as animal feed to cattle feed lots.

The vegetative development of a sunflower is divided into two phases: *vegetative emergence* and *vegetative development.*

Vegetative emergence covers the period from seedling emergence to when the first true leaf is less than 4 cm long.

The vegetative development period nomenclature is based on the number of true leaves over 4 cm in length. For example, if there are two leaves over 4 cm, the growth stage is described as V2; if there are four leaves, the stage is V4.

The reproductive phase is separated into nine stages based on flower development:

- Stages R1 to R4 describe the stages from when the flower bud first emerges through to just before the start of flowering.

- Stage R5 describes the beginning of flowering and is sub divided to describe the percent of the flower that has completed or is in flower such as R5.1 (10%), R5.5 (50%), R5.9 (90%).

- Stages R6 to R9 cover the period from when flowering is complete (R6) through to physiological maturity R9.

Sunflowers are known as *composite flowers*. The large flower head at the top of the plant is a composite of hundreds of small disc flowers. The male (stamen) and female (stigma) are both present in the disc flowers. These individual flowers are what ultimately produce seeds or, as we commonly call them, sunflower seeds.

The bright yellow petals on the outer circle of the sunflower head are called *ray flowers*. Sunflowers depend largely on bees and other winged creatures to spread the pollen from the stamen to the stigma. The brightly coloured ray flowers attract pollinators to the disk flowers.

Sunflowers exhibit *heliotropic movement* during the daytime. The stems will bend and the sunflower head will tilt west to face towards the afternoon Sun. At night-time, the stem will return to being fully erect. As the early morning Sun rises in the east, the sunflower head will tilt east. [53][54][55]

Figure 5-13 illustrates the sunflower structural features.

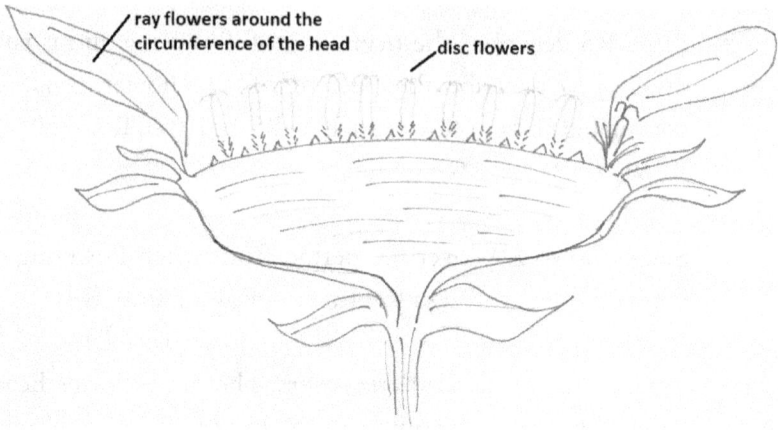

Figure 5-13
Sunflower structure

Sugar Beet *(Beta vulgaris subsp. altissima)*

- Seed Structure: Dicot
- Growth: Epigeal
- Growth Period: up to 150-200 days

Sugar beet traces its origin to the Middle East and Mediterranean regions. Moisture (irrigation) is essential along with heat when growing beet crops. In the US, growing sugar beet as a crop for obtaining sugar began in earnest in the 1870s. Today, sugar beets are grown in 11 US states. About 55% of US sugar production derives from beets with 45% coming from sugar cane. In Canada, about 37,000 acres of sugar beets are planted each year with 1/3 of these acres being in Ontario and the balance in southern Alberta near the town of Taber.

As a curious aside, many Canadians grow Swiss Chard in their gardens each year. This hardy plant which seems to enjoy heat and moisture is the *flavescens* sub-species of *Beta vulgaris*. Gardeners sometimes refer to Chard as leaf beet.

Sugar beet is a biennial crop. In the first year of growth the plant develops leaves and a beet. In the second year the plant goes on to express flowers and seeds. Sugar beet growers have little interest in flowers and seeds and so sugar beets are grown and harvested annually for the beet. Each root (beet) contains approximately 75% water, 20% sugar and 5% pulp.

At a sugar beet processing plant, the beets are chopped into finer pieces and exposed to hot water in a processing vessel. The heat causes molecules of sugar to solubilize from the beet material and to precipitate out. The molecules of sugar are collected for further processing. The sludge remaining in the process vessel is called *molasses*. The sugar recovery process is only about 87% efficient, so the remaining molasses sludge will have a residual sugar content. The molasses and pulp are sold to cattle feedlots.

As part of my recipe development and consulting efforts with distillers, I have made distillate from sugar beet molasses. After barrel ageing the distillate for a while, the resulting product was on par with any of the big-name commercial rum varieties that consumers like to mix with cola. However, I advise craft distillers to focus on molasses from sugar cane. The flavour profile of rum made from sugar cane molasses is superior in taste. This is related to the fact that the sugar content of sugar beet is mainly sucrose. The sugars in sugar cane are more complex structures and deliver a more complex taste profile.

Optimal sugar beet growth requires a dense leaf canopy. Sugar beets are thus planted as early as feasible in the spring with the understanding that the young plants are temperature and moisture sensitive.

About three to four weeks after emergence, the plant will have developed 6 leaves. At this point, rapid vegetative growth will commence. The plant will gain in height and the leaf canopy will expand. When the canopy of one plant overlaps with an adjacent plant, the canopy will have a Leaf Area Index (LAI) of 3, meaning the canopy area is 3 times the ground area at the plant base. With a LAI of 3, the plant will utilize near 90% of solar light for photosynthesis and development of the beet.

The sugar beet plant grows until harvested or growth is stopped by a hard freeze (minus 5°C or more). Sugar beets primarily grow tops until the leaf canopy completely covers the soil surface in a field. This normally takes 70 to 90 days from planting. Thereafter, growth is contained to the beet beneath the surface. [56][57][58]

To Sum It Up

» The visible light spectrum covers the range 400 to 700 nanometers.

» Plant leaves contain *phytochrome* and *cryptochrome* protein substances. Under the influence of red light (red light has a wavelength of 700 nanometers) phytochrome will trigger vegetative growth. Under the influence of far-red light (far-red light has a wavelength of 780-800 nanometers), vegetative growth will diminish and eventually cease. Cryptochrome proteins react similarly to changes in blue light and UV-A light. Blue light has a wavelength of 450-495 nanometers. UV-A light with a wavelength of 315-400 nanometers.

» Plants are generally divided into *long-day* and *short-day* sub-categories.

» Long-day plants will start to flower (reproductive growth) when darkness hours are at or near a minimum (around the Summer Solstice). Short-day plants will start to flower when dark hours exceed a minimum threshold (late in August).

» Flax dates to 9000 years ago in Turkey where it was used to obtain fibers and oil. Flax oil is high in healthy, unsaturated fatty acids.

» Canola is part of the genetic family that includes broccoli, cauliflower, cabbage, kale, bok choy, arugula, Brussels sprouts, collards, watercress and radishes. The developmental history of canola starts in the WW II era, with Canada leading the way.

» Mustard is similar in growth patterns to canola. Main varietals are yellow mustard (*Sinapsis alba*) used in prepared mustards, and brown/oriental mustard (*Brassica juncea*) used in Dijon mustards.

» A legume plant is capable of interacting with *Rhizobia* bacteria in the soil to fix Nitrogen into the sub-surface soil. Nitrogen added to soil, whether by Nitrogen fixation or by chemical fertilizer, is beneficial to the next year's crop.

» Lentils are a legume that early Neolithic man recognized after the Younger Dryas period. Peas were also recognized in this same era (c10,000 B.C.) in the Fertile Crescent region. The chick pea is a legume that dates to 9500 years ago in the southeast of Turkey.

» Corn was already growing in North America at the time of Columbus. It is thought that corn traces its lineage to a seed-bearing grass species called *teosinte* that arrived in North America via an ancient land bridge that once connected Asia to North America in an area that today is the Bering Sea.

» Cereal grains are named after Ceres, the goddess of crops and agriculture, in ancient Roman society. Emmer and Einkorn are two of the earliest varietals of cereal.

» Soybean cultivation is thought to have started in China about 3000 years ago. In the 1600s, the plant was introduced to Europe. The soybean plant was brought to the British Colony of Georgia in the mid-1700s. In the mid-1900s, the plant was introduced to South America.

» Prior to the late 1990s, canary seed was only suited for bird consumption owing to the presence of hair-like fibers called trichomes which are attached to the seed hulls. Thanks to plant breeding efforts at the University of Saskatchewan, hairless canary seed varietals were created. These are fit for human consumption and appear to have dietary health benefits.

» As plant based meat substitutes gain in popularity, expect to hear more about Faba beans, which have protein content >25% , and >40% carbohydrate content.

» Buckwheat is a pseudo cereal. At one time, the primary end market for buckwheat was Japan where it was used in soba noodles. Demand for buckwheat in North America is now increasing as consumers seek out gluten-free, nutritious, ancient grains.

» It is estimated that the early residents of northern Mexico domesticated the sunflower around 3000 B.C. Today, Russia and Ukraine account for 50% of global sunflower output. About 75% of the sunflowers grown in the US are grown in the Dakotas and Minnesota. In Canada, the province of Manitoba is the focal point for sunflower production. Owing to their oil content, sunflower prices tend to track soybean futures prices.

» Sugar beet traces its origin to the Middle east and Mediterranean regions. About 55% of US sugar production derives from beets.

CHAPTER 6
PHOTOSYNTHESIS

Photosynthesis is the sunlight-driven, metabolic process that drives plant growth. In the photosynthetic process, the plant leaf tissue takes in carbon dioxide (CO_2), moisture (H_2O), and light energy from the Sun. The CO_2 is transformed to a more energy efficient format, the plant tissues exhibit growth and the plant expels oxygen (O_2). What follows in this chapter is a step-by-step overview of how Nature uses sunlight to grow plants.

At the heart of the photosynthetic process are three items: an enzyme called *RuBisCo*, a substance called *ATP*, and a substance called *NADPH*.

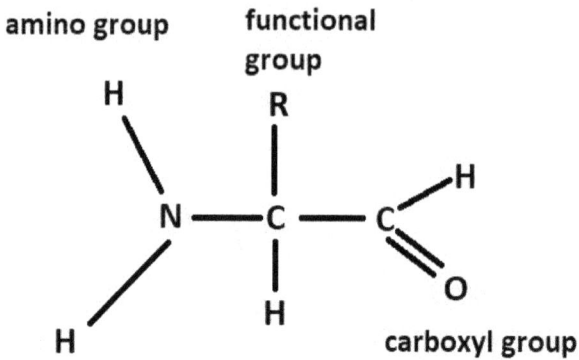

Figure 6-1
Protein Structure

An enzyme is a protein structure, the skeleton of which is shown in Figure 6-1. At one end of the protein structure is an H-N-H amino group. At the other end is a O=C-H carboxyl group. Attached to the molecule is a functional "R" group which can be a very complex structure. In the photosynthetic process, the critical enzyme that makes the reaction run is ribulose-1,5-bisphosphate carboxylase-oxygenase. For simplicity, it is nicknamed RuBisCo.

Adenosine Tri-Phosphate, or ATP, is a complex molecule that captures cellular energy and releases it when and as required to drive cellular reactions. Think of ATP as a bank where energy is deposited and then withdrawn as required.

NADPH (Dihydronicotinamide-Adenine Dinucleotide Phosphate) is the molecule NADP that has gained a Hydrogen atom (and thus one electron). NADPH is fundamental in making cellular reactions run, just like the wheels on a tractor are fundamental in making that tractor move forward.

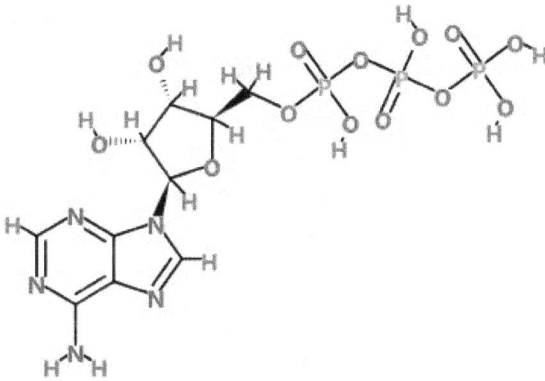

Figure 6-2
ATP Structure

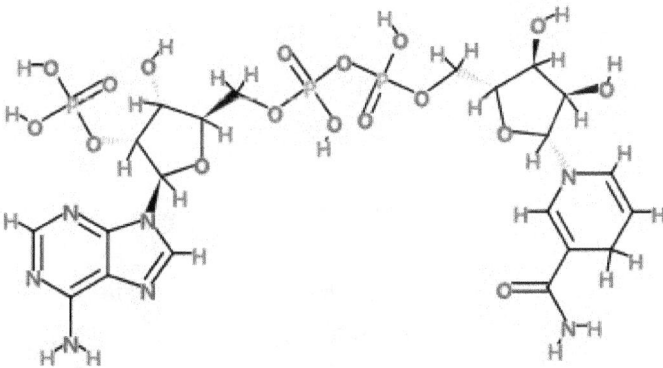

Figure 6-3
NADPH Structure

Leaf Cellular Structure

The cell structure of a plant leaf comprises a primary cell wall and a secondary cell wall. The primary cell wall structure provides support to the leaf during growth. The secondary cell wall is located inside of the primary cell wall to provide support after the leaf stops growing. The primary cell wall is built mostly of *cellulose* which is a chain-like structure of glucose $(C_6H_{12}O_6)$ molecules. The cellulose molecules are assembled into fibers called *microfibrils* which provide the mechanical strength to the cell wall. Cell walls also contain *hemicellulose* and *pectins*. Hemicellulose is a chain-like structure comprised of 5-carbon variants of glucose. In the leaf cell structure, hemicellulose chains are cross-linked to the cellulose chains. Pectins help to bind together individual cells and also serve to regulate water uptake by the cells.[1] The surface of a plant leaf contains pores called *stomata* which allow the leaf to take in CO_2 and expel O_2.

The cells in the middle of the leaf structure are called *mesophyll cells*. Each mesophyll cell contains smaller structures called *chloroplasts*. Inside each chloroplast are structures called *thylakoids* which contain a green-colored pigment called *chlorophyll* which is capable of absorbing light. The chloroplast structures are where photosynthesis reactions occur.

Photosynthesis can be divided into two stages: the light dependent reactions and the light independent reactions (*Calvin Cycle*). These two stages are dependent on one another and work simultaneously.

As light energy from the Sun hits the chlorophyll in the thylakoid portion of the mesophyll cells, a residual phosphorous atom (P*i*) present in the leaf tissue helps in the conversion of the photon light energy to NADPH and ATP chemical energy. The moisture (H_2O) absorbed by the plant is stripped of its H atoms which combine with NADP to make the NADPH. The remaining Oxygen atom is expelled into the atmosphere. This is how plant life creates the Oxygen that we breathe.

Meanwhile, in the stomata cells in the surface of the leaf tissue, the Calvin Cycle takes in CO_2 and with energy from ATP and NADPH plus help from a 5-carbon acceptor molecule, ribulose-1,5-bisphosphate (RuBP), makes a 6-carbon compound. This 6-carbon molecule is unstable and with a bit of help from the enzyme ribulose-1,5-bisphosphate carboxylase-oxygenase (RuBisCo), produces 3-carbon sugar molecules called 3-phosphoglyceric acid (3PGA).

Next, ATP and NADPH convert the 3PGA into a 3-carbon sugar molecule called glyceraldehyde-3-phosphate (G3P).

Next, the Calvin Cycle takes in additional CO_2 molecules which are used to create six more G3P molecules. One of these will exit the reaction process and evolve into the 6-carbon sugar glucose. The glucose and its associated formats (hemi-cellulose, beta glucan) are used to help the growing plant create more stalk material and endosperm starch material. The other five G3P molecules will be used to regenerate more RuBP carbon acceptor molecules so that the Calvin Cycle can keep working over and over again.

Figure 6-4 illustrates a summary of the overall photosynthetic process.

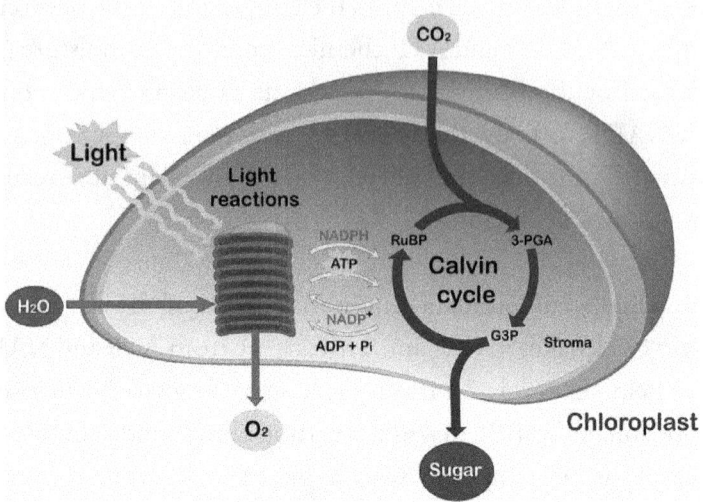

Figure 6-4
Overall Photosynthesis Process

CHAPTER 7

HAIL DAMAGE AND INSURANCE COMPENSATION

To financially protect against having crops damaged by hailstorms, many farm operators each year purchase hail insurance. In Canada, crop hail insurance plans vary in structure. In some provinces, hail insurance is part of all-peril crop insurance plans. In other provinces, there are dedicated insurance providers that specialize in offering hail insurance. In the USA, there are a variety of private hail insurance companies.

In both countries, the insurance industry is a complexity of varying policy types, and indemnification policies. For example, an insurance company offering hail insurance (or crop insurance with hail as a listed peril) might be financially backed by a larger global re-insurance company. The re-insurer will have strict guidelines as to operating and financial ratios that must be generated by the insurance company over time. This re-insurance model is not limited to just hail insurance providers. The insurance company that has sold you a home insurance policy is also backed by a larger re-insurer who has strict guidelines. As an example, I was recently told that if I made a claim against my home insurance policy, my annual premiums would increase by 15% each year for three years. In other words, the re-insurer organization really does not want people making claims unless absolutely necessary. Thankfully, this degree of rigidity has not made itself apparent in the hail insurance part of the insurance spectrum.

Re-insurers generally offer two types of re-insurance: *Treaty* and *Facultative*. For hail insurers, the Treaty method will apply in that the re-insurer agrees to back all the policies written in a given year for a specific type of insurance. Re-insurance may be considered *proportional* or *non-proportional*. Under the proportional model the re-insurance firm receives a percentage share of all policy premiums sold by the hail insurance company. If a hail claim is recorded by a farm operator, the re-insurance firm bears a portion of the losses based on a pre-negotiated percentage. With non-proportional re-insurance, the re-insurance firm is liable if the hail insurance company experiences losses that exceed a specified amount. This amount is known as the *retention limit*.

A hail insurance policy will carry a *premium* that the farm operator must pay. Different insurance firms will impose different premium levels. These levels will be based on the past hail experience in the jurisdiction where the farm operator's land is located. The premiums will also vary according to the type of crop. The following three examples from various Canadian and American hail insurance providers were easily found in an Internet search and are not particular to any one insurance provider. Policy structures across North America are not necessarily limited to just these three example types.

Full Coverage Policies

One style of policy is the *full coverage* policy which provides for a set amount of money per acre depending on the percent damage to the crop. For example, consider a farm operator who has $300 per acre 'full coverage' on his wheat crop. He encounters a hail storm and the adjustor determines the crop has sustained 20% damage. The farm operator stands to receive a financial settlement of $300 x 0.20 = $60 per acre, less any policy premiums paid to acquire the insurance. The table in Figurer 7-1 illustrates further:

Hail Loss Identified by Adjustor	Payable Losses to Farm Operator
10%	10%
20%	20%
30%	30%
40%	40%
50%	50%
60%	60%
70%	70%
80%	80%
90%	90%
100%	100%

Figure 7-1

The table in Figure 7-1 shows that a 100% loss identified by the adjustor will result in 100% of the per-acre insured amount to be paid out. Various insurance companies will have a loss level that triggers a 100% payout. When taking out hail insurance, a farm operator should clarify with the insurance company what their exact trigger level is for a full loss.

Straight Deductible Policies

Another policy variation is the *straight deductible*. The payable losses arising from hail damage will be reduced by the deductible amount stated in the policy. The advantage to the farm operator using this type of policy is a slightly cheaper premium cost for obtaining the policy. The premium cost is cheaper because the farm operator is willing to accept a lesser payout at various crop damage levels. For example, consider a farm operator who has $200 per acre coverage on his canola crop along with a '20 straight deductible' clause.

Hail Loss Identified by Adjustor	Payable Losses to Farm Operator
10%	0%
20%	0%
30%	10%
40%	20%
50%	30%
60%	40%
70%	50%
80%	60%
90%	70%
100%	80%

Figure 7-2

The table in Figure 7-2 shows the payable losses that would apply in this type of policy. Note that a hail insurance adjustor would have to identify greater than a 20% loss level before the farm operator receives payment.

Disappearing Deductible Policies

Another policy variation is the disappearing deductible. For example, suppose a farm operator takes out $150 per acre coverage on his flax crop along with a '10 disappearing deductible'. The farm operator will only receive a financial settlement if the adjustor assigns 10% or greater loss to the hailed crop. As the amount of hail damage increases, the deductible is reduced on a sliding scale. The advantage to the farm operator with this type of policy is a reduced policy premium. The following table shows the payable losses with this type of policy. More accurately, this policy would be described as a '10 disappearing deductible', disappearing at 30.

Hail Loss Identified by Adjustor	Payable Losses to Farm Operator
10%	0%
20%	10%
21%	12%
22%	14%
23%	16%
24%	18%
25%	20%
26%	22%
27%	24%
28%	26%
29%	28%
30%	30%

Figure 7-3

As the table in Figure 7-3 shows, once the adjusted damage reaches the 30% level there are no further reductions to the payable losses.

Different insurance companies will have varied nuances on these policy types. Farm operators should exercise due diligence and decide carefully what type of policy best suits their financial needs.

For jurisdictions in Canada where hail is part of an all-peril type policy, it is common for the federal government to also play a role in determining the policy premiums. In these jurisdictions, it is further likely that the insurance policy will also factor in the expected economic value of the crop being grown. These types of policies are outside the scope of this chapter.

In my travels adjusting crops for hail damage, the one common thread I have observed is a lack of adequate insurance. Carrying $25 per acre of hail insurance provides little benefit to any farm operator.

If you think of hail insurance in the context of personal insurance, would you ever knowingly reduce the amount of fire insurance on your house to save money on the premiums? Would you ever reduce the amount of liability insurance on your vehicle policy because you thought your chances of causing 3rd party injury or loss were low? Accidents happen. Hail storms happen.

As one farm operator bluntly expressed it to me, *"Hail insurance premiums are just another input along with seed, fuel for the tractor, fertilizer and pesticide chemical. Budget accordingly and make sure you are fully insured"*.

In many ways, the procedure for adjusting hail damage is no

different than it would be for a house or car that receives hail damage during a storm. An insurance adjustor examining a vehicle with hail damage will quickly assess the number of hail dents on the vehicle, their size and their depth. The adjustor will refer to a set of vehicular indemnity tables that reflect the cost for an auto body shop to complete the repairs to the vehicle.

The key difference between hail damage to a fixed object such as a car and hail damage to an agricultural crop is the fact that the crop, despite the hail damage, has the potential to continue growing if the hail damage is incurred early in the growing season. A damaged car is not a living organism. It will not heal the dents on its roof and restore itself.

The job of the crop hail insurance adjustor then is complicated. Not only does the adjustor have to accurately assess the extent of physical damage to the crop, but must also accurately identify the stage of growth the field was at when it was hailed upon. This critical determination will determine how the indemnity tables calculate the *payable loss* figure. Furthermore, the adjustor must be able to discern between hail damage to the plant versus insect damage, wind damage, and even damage pertaining to a hot, dry growing season.

What follows in this chapter is an overview of how hail can impart damage to crops. The information in this chapter has been taken from plant defoliation academic studies done by universities around the world. This chapter will be helpful to the farm operator who is seeking to gauge the damage to his crop in the aftermath of a hail storm. This chapter might also be helpful for a trainee-adjustor who is preparing to write the licensing exam.

Flax *(Linum angustafolium)*

An early pioneer in the study of damage to flax plants was Idaho crop scientist K.H.W. Klages. In his research, he simulated hail by inflicting mechanical damage to flax plants using a branch of white willow as a whip. He also used a pair of scissors to cut flax plant stems to further simulate hail damage. In 1933 he published his research findings in the *Agronomy Journal.* (1)

Flax is an epigeal germinator. The cotyledon tissue is pushed above the ground by the hypocotyl tissue. If hail cuts off the flax plant *below* the cotyledon leaves, the plant will not recover. Klages focused on cutting the plant just above the cotyledon tissue. His conclusion was that if hail damages the plant so bad that only one cotyledon leaf remains, the plant will still survive, but its yield will be negatively affected.

Klages continued his work by cutting plant stems after six inches of growth and after nine inches of growth. His conclusion was that flax stems at these vegetative stages cut by hail will not survive.

He next studied the effect of defoliation during vegetative growth. His conclusion was that a flax plant defoliated by a hail storm during vegetative growth will recover and recuperate.

He went on to study the effect of hail on plants that were approaching the bloom stage. He used a willow branch to physically whip the plants. He determined that a whipped plant at a growth stage up to the bloom stage will exhibit a fair degree of recovery.

He next turned his focus to flax plants that had flowered and expressed bolls. He concluded that his whipping action was detrimental to yields. He further noted that the degree of recovery was related to climatic conditions in the days following the damage.

The stage of growth combined with the implications of Klage's 1933 work form the basis for payable losses offered by hail insurance companies to flax growers. Some of Klage's work was repeated and validated at the University of Minnesota in 1970.[2] In 1999, an Italian study confirmed that a stand of flax plants is capable of responding to a decrease in plant stand density by expressing more bolls per plant.

The key message in the context of indemnity awards for hail-damaged flax is that flax plants damaged in the early stages of growth are resilient. It is only when hail damage occurs after boll formation that significant loss can result. But even then, the bolls are resilient and hard to dislodge from the plant.[3] Flax is also prone to disease damage, for which hail insurance offers no compensation.

Diseases that can affect flax include: wilt, rust, pasmo, and mildew.

- Wilt is caused by a soil-borne fungus which impedes the plant's ability to uptake water through the stem tissues. The plant will appear yellowish due to the moisture starvation.

- Rust is a significant threat to flax and is caused by the soil-borne bacteria *Melampsora lini*. Orange-colored pimples on leaves are indicative of rust.

- Pasmo is caused by the soil-borne *Septoria linicola* fungus. Brown lesions will be evident on leaves and stems. Pasmo can lead to weakened boll stem structures and boll losses due to wind and rain which are associated with hail storms.

- Mildew is caused by the fungus *Oidium lini*. The obvious sign of mildew is a white, powdery substance on the leaves of the plant.

Canola *(Brassica napa* and *Brassica rapa)*

In 1987, Agriculture Canada plant scientist D.I. McGregor published his findings of a study designed to investigate the effect of hail on canola during the early vegetative stages of growth. This work has been incorporated into the crop hail insurance adjustment methodology. As McGregor specifically states in his report:

Adjusting for hail injury to a rapeseed crop to settle an insurance claim is a process of predicting the loss in harvestable seed. In addition to assessing the extent of injury to the plants, the ability of the crop to recover and compensate for such injury must be taken into account.

What McGregor determined was that thinning the plant stand density encouraged surrounding plants to grow larger. McGregor further concluded that hail damage at an early stage will impair crop yield by greater than 20% if and only if plant density has been reduced to under 40 surviving plants per square meter. [4]

In 2016, the results of a five-year joint study between the University of Saskatchewan and National Crop Insurance Services (NCIS) were published. [5] In this work, canola plants were 100% and 50%

defoliated at five leaf stages, namely stages: 2, 4, 6, 8, and 10. The yields of these damaged test plots were compared to an undamaged control group.

The researchers averaged the data across the five stages of growth. Observations were:

- The 50% defoliated plants had a seed yield of 77 to 97% (average 92%) of the control test plots.

- The 100% defoliated plants had a seed yield of 68 to 81% (average 75%) of the control test plots.

Looking specifically at the results of the 2-leaf and 10-leaf damaged plants provides further insight.

- 100% defoliation at the 2-leaf stage resulted in a yield that was 86% of the control plot yield.

- 100% defoliation at the 10-leaf stage resulted in a yield that was 66% of the control plot yield.

- 50% defoliation at the 2-leaf stage resulted in a yield that was equal to the control plot yield.

- 50% defoliation at the 10-leaf stage resulted in a yield that was 80% of the control plot yield.

The key point of this work is that the canola plant is a resilient plant. When injured by defoliation, more sunlight will penetrate the canopy and photosynthesis will provide energy for continued growth. Carbohydrates stored in stems and roots will be remobilized to provide nutrition for recovery growth.

A 2019 study at the University of North Dakota added more data to the canola resiliency argument. In this study, four growth stages (4th-5th leaf rosette, bolting, 50% flowered, 90% flowered) and four stand reduction levels (25%, 50%, 75%, 90%) were used. When the results were averaged across the four levels of reduction, it was revealed that:

- Damage at the 4th-5th leaf rosette and the bolting stage resulted in a 21% reduction in yield as compared to the undamaged control plants.

- Damage at the 50% flowered level showed a 28% reduction in yield as compared to the control plants.

- Damage at the 90% flowered level showed a 46% reduction in yield as compared to the control plants.

The key point is that canola will suffer its greatest yield reduction when damaged late in the vegetative growth stage.

This study went on to explore stem cut-off, as might be expected from a severe hail event. Four growth stages (4th-5th leaf rosette, bolting, 50% flowered, 90% flowered) and four cut-off amounts (25%, 50%, 75%, 90%) were studied. The plants were cut at one-half their height. When the results were averaged across the four levels of reduction, it was revealed that:

- Damage at the 4th-5th leaf rosette and the bolting stage

resulted in a 6% reduction in yield as compared to the undamaged control plants.

- Damage at the bolting stage resulted in a 3% reduction in yield as compared to the undamaged control plants.
- Damage at the 50% flowered level showed a 5% reduction in yield as compared to the control plants.
- Damage at the 90% flowered level showed a 22% reduction in yield as compared to the control plants.

The key observation is that canola will suffer its greatest yield reduction when stems are cut late in the vegetative growth stage.[6]

The hail insurance industry has taken McGregor's work into account as well as all work subsequent to McGregor when calculating payable losses after a hail event. The canola plant in the vegetative growth stages is resilient and indemnity awards will be structured accordingly.

Where the canola plant loses its resilience is in the podding and physiological maturity stages. A plant damaged in the reproductive stages cannot express much recovery. Hail adjustors will base their conclusions on a sample of plants that display pod losses due to hail. Indemnity awards will be calculated accordingly. For late-stage canola hailed while standing, this sampling is straightforward. Late-stage canola hailed in the swath will require a bit more investigative work as not all parts of the swath will have been uniformly damaged. Different insurance companies will have different approaches to dealing with canola damaged in the swath.

Canola is also prone to disease and insect damage, for which hail insurance offers no compensation. Diseases and damage include:

- Blackleg: a fungal infection from spores of *L. maculans* and *L. biglobosa*. Canola with blackleg will display black specks on leaves. As infection grows, the plant stems will exhibit black discoloration.

- Stem rot: the fungus *Sclerotinia sclerotiorum* is responsible for stem rot. The appearance of grey-white discoloration at stem branch points is evidence of stem rot.

- Fusarium wilt: the spore *Fusarium oxysporum* is responsible for wilt. The spores damage the plant vascular tissue and moisture uptake is impeded causing the plant to wilt.

- Clubroot: soil-borne spores of the fungus *Plasmodiophora brassicae* are responsible for clubroot. (These spores can be transferred long distances on vehicles, so an adjustor should obtain definitive permission from the farm operator before driving a vehicle into a harvested canola field).

- White rust: some varietals of canola can be affected by spores of *Albugo candida*. The canola stems and branches can take on a whitish, blistered appearance after a rainstorm. The blisters can almost resemble whitish hail damage marks.

- Grasshoppers: grasshopper damage on pods must not be mistaken for hail damage to pods. Grasshopper damage is characterized by small holes that appear to have been chewed through the pods.

Mustard *(Sinapsis alba* and *Brassica juncea)*

Mustard is generally adjusted in a like manner to canola. Many of the same airborne and soilborne pathogens that affect canola can also affect mustard varietals.

Lentil *(Lens culinaris)*

In 2011, University of Saskatchewan plant scientist Rosalind Bueckert published the results of her 2006 and 2007 studies on large green lentils (*CDC Sedley* varietal) and red lentils (*CDC Blaze* varietal). Plants were manually damaged using a flail. The flail was made of a 1.5-meter-long wooden handle; attached perpendicular to the handle was a 0.6-m x 50-mm wide section of angle iron, and hooked to the angle iron were six rubber bungee cords and four strings of glass beads, all 0.4 m long.

One pass with the flail was deemed to have imposed 30% damage to the plants. Two passes inflicted 60% damage and three passes inflicted 90% damage.

Damage was imposed to the plants at four stages: vegetative growth, early flowering, pod filling, and physiological maturity.

When all levels of damage were averaged across the four growth stages, results showed:

- Vegetative stage growth suffered a 28% yield loss as compared to the undamaged control plants.

- Early flower stage growth suffered a 37% yield loss as compared to the undamaged control plants.

- Podding stage growth suffered a 45% yield loss as compared to the undamaged control plants.

- Physiologically mature plants suffered a 45% yield loss as compared to the undamaged control plants.

The key point is that as lentils advance towards maturity, the damage from a hail storm will impose greater yield loss damage.

When just the 90% level of damage was averaged across all four growth stages, the results showed:

- Vegetative stage growth suffered a 53% yield loss as compared to the undamaged control plants. (range was 35% to 71%)

- Early flower stage growth suffered a 66% yield loss as compared to the undamaged control plants. (range was 61% to 70%)

- Podding stage growth suffered a 66% yield loss as compared to the undamaged control plants. (range was 62% to 69%)

- Physiologically mature plants suffered a 66% yield loss as compared to the undamaged control plants. (range was 56% to 85%)

The deviation range in these damage levels is explained by the fact this work was done over two seasons, using two locations, and two varietals. The observation is that greater damage will occur on more mature lentil plants.

The lentil plant is resilient and displays indeterminate growth. In the hours after a hail event, the lentil crop will probably look beaten to death. By the time a hail adjustor arrives on site, the crop will have recovered somewhat and will not look as bad.

The argument that I heard over and again when adjusting lentils in 2020 and 2021 revolved around the sensitive nature of the plant. Farm operators stated loudly that the hail event had stressed the plants and their potential yield had been harmed. The excellent work by Bueckert and her research colleagues has certainly quantified this argument. [7][8]

Lentils are also prone to disease, for which hail insurance offers no compensation. Diseases include:

- Ascochyta Blight: this is caused by spores from *Ascochyta pinodes*, *Ascochyta pinodella*, and *Ascochyta pisi*. This blight is evidenced by grey to tan spots or lesions (with dark margins) on leaflets, stems, flowers and pods. The lower leaflets are the first to be affected and the spores then work their way up the plant.

- Anthracnose: this condition is caused by fungi of the genus *Colletotrichum*. It appears as white to grey or cream-colored spots on leaflets and stems. As the condition spreads, areas of the field will appear yellow as plants die off.

Peas *(Pisum sativum)*

Peas do not appear very often as the subject of a simulated hail damage study. The one study that remains prominent is the

1984 effort by University of Montana researchers Miller and Muehlbauer.[9] Their work involved the *Alaska* varietal of pea and three stages of damage.

When sample plants were at the 8th node of vegetative growth, the stems were incised between the 4th and 5th node. (Treatment 1)

When other sample plants were at the bloom stage, they were incised between the first and second flowering node. (Treatment 2)

When yet other sample plants were at the bloom stage, they were incised below the first flowering node. (Treatment 3)

For Treatment 1 and Treatment 2 damage, results showed that the pea plants exhibited similar pods per plant and peas per pod as compared to undamaged control plants.

For Treatment 3 damage, results showed a significant reduction of peas per pod, but a similar number of pods per plant as compared to undamaged control plants.

For Treatment 1, pea yield will drop 0.41% for each 1% increase in plant damage.

For Treatment 2, pea yield will drop 0.59% for each 1% increase in plant damage.

For Treatment 2, pea yield will drop 0.85% for each 1% increase in plant damage.

The takeaway from this work is that the pea plant can exhibit recovery. However, maturity will be delayed, plant height will be reduced, the extent of branching will be expanded, and pea weight will be reduced. Hail insurance providers will structure their indemnity awards relying to some extent on this work by Miller and Muehlbauer.

Peas are also prone to disease and insect damage, for which hail insurance offers no compensation. Diseases can include:

- Root Rot: this is caused by the bacteria *Fusarium solani*. Red to reddish-brown lesions on roots and stems are a sign of this disease.

- Ascochyta Blight: this is caused by spores from *Ascochyta pinodes*, *Ascochyta pinodella*, and *Ascochyta pisi*. This blight is evidenced by grey to tan spots or lesions (with dark margins) on leaflets, stems, flowers and pods. The lower leaflets are the first to be affected and the spores then work their way up the plant.

- Sclerotina: the fungus *Sclerotinia sclerotiorum* is responsible for stem rot. The appearance of grey-white discoloration at stem branch points is evidence of stem rot.

Chick Peas *(Cicer arietinum)*

Chick peas, much like peas, do not appear very often as the subject of a simulated hail damage study. One study that appears in the literature is the 2004-2005 joint effort between crop scientists at the University of Saskatchewan and Guelph University in Ontario.[10]

This study focused on growth plots near Saskatoon and Swift Current, Saskatchewan. Two chick pea varietals were studied: *Sanford* and *CDC Yuma*. Plant seeding density was about 65 plants per square meter.

Some of the study plants were 50% defoliated during vegetative growth and others were 50% defoliated when half of the plants were exhibiting first flowers.

The interesting outcome of this work is that the defoliated plants exhibited only a marginal difference in yield as compared to undamaged control plants. As the report states:

Even with defoliation treatments, chickpea had the ability and sufficient time to replace lost leaves and compensate for lost metabolites from the removal of leaves. As a result, yields from defoliated plants were similar to the control treatment.

The takeaway from this work is that the chick pea plant can exhibit recovery. The hail insurance providers will structure their indemnity awards relying to some extent on this work.

Corn *(Zea mays)*

A 2019 published paper by crop scientists at Virginia Tech does an excellent job of summarizing efforts at understanding hail damage to corn crops. As the authors point out, hail can damage young plants and create stand reduction. Hail can shred leaf blades and reduce the available surface area for photosynthesis. Hailstones can bruise plants and increase risks of plant stalk breakage. Hail striking plants at the time of pollination can destroy tassels and silks. Late season hail events can lead to ear loss or bruised kernels. Simulating hail damage to corn crops is thus not an easy task. Simulation work has been further complicated as new corn hybrid varietals have entered the marketplace. Many of these exhibit different responses to damage as opposed to older, more traditional varietals.

Defoliation Damage up to Stage V10

Studies over the past 50 years have indicated that the effect of defoliation damage in early vegetative growth stages varies from slight to not at all. The insurance industry has taken note of this work and indemnity awards are structured accordingly.

Defoliation Damage after Stage V10

Reported studies in the literature for post-V10 growth vary markedly. The most recent study was a 2005 effort that concluded for stages V14 to R3, corn yield will decrease linearly as the amount of defoliation increases.

Insurance Tables

Crop adjustors often use what is called the "hail adjustor's horizontal leaf method" to quickly determine the corn growth stage at which hail damage has occurred. Vegetative stages in this scale are determined by counting the number of leaves whose tips are bending over from horizontal. Once the growth stage has been identified, insurance providers then resort to tables to estimate the economic yield losses due to defoliation.

For example, damage up to stage V10 might provide the farm

operator with practically no indemnity losses. Damage at the V16 stage might trigger just under 20% indemnity losses only once 40% leaf area damage has been incurred by the corn crop. Different insurance providers will take different approaches to calculating loss compensation.[11]

Damage to Stalks and Ears

There is a scarcity of academic literature focused on simulated damage to stalks and ears. One paper that does provide some insight was written in 1935 by crop scientists at Iowa State College.[12] The stalks and ears were bruised by striking each stalk five blows and each ear two blows with an instrument comprising a paddle-like handle about 20 inches long to which was fastened a hardwood knob about ½ inch in diameter. The blows, which were distributed uniformly over the lower two-thirds of the plant, were struck hard enough that at least one or two of them caused the knob to sink into the stalk, producing a bruise similar to that from a medium sized hail stone striking the stalk with great force. The two blows on the ear were struck with sufficient force to penetrate the husks and bruise the kernels. Results of this work showed:

- Severe bruising of stalks and ears a week before the tasseling stage reduced the yield 20 percent.

- Severe bruising of stalks and ears at the tasseling stage reduced the yield 35 percent.

- Severe bruising of stalks and ears and severe shredding of leaves a week before tasseling reduced yields 64 percent.

- Severe bruising of stalks and ears and severe shredding of leaves at the tasseling stage reduced yields 77 percent.

Silage Corn and Seed Corn

Literature studies generally do not distinguish between corn grown for seed and corn grown for silage. A 2008 study did, however, shed insight into the silage issue. Corn plants at the V7, V10, R1, and R4 stages were all subjected to controlled levels of defoliation. The plots were harvested at the 50% milk stage. Wet chemistry analysis was performed on the silage samples to determine protein, and dry matter fiber contents. This data was input into existing mathematical models to estimate dairy animal milk production, and digestibility. Milk production estimates were based on a standard dairy cow with 613 kg of body weight producing 36 kg of milk per day at 3.8% fat. Results showed that the greatest negative changes to wet chemistry properties were from plant material damaged at the R1 and R4 stages.

The hail insurance industry has likely incorporated this study into its indemnity tabulations. For corn grown as silage, indemnity tables will differ from those used for corn grown for seed. The reasoning behind this is with silage corn, even though the plants might be hail damaged, the corn cob will provide nutrition to the animal consuming the silage. [13]

Corn is a difficult crop to hail adjust. Farm operators should make every effort to cooperate with their adjustor as he applies the latest insurance industry procedures.

Wheat, Barley, Rye, Oat and other Cereals

When adjusting cereal grains, the primary focus of the adjustor will be to determine the stage of crop growth at the time of hail. The North American hail insurance industry regards cereal growth stages as being: Grass (Emergence) to Jointing, Jointing to Boot, Boot and Just Headed, Bloom, Milk, Soft Dough, Hard Dough, and Combine Ripe.

A 1937 study at Iowa State College seems to have been the first work designed to examine regrowth potential of damaged oats, barley, and wheat. In this study hail was simulated by the use of a hand-held whip. [14]

Results indicated:

- Grain severely injured in the grass stage is reduced in yield 10 to 50 percent, depending on growing conditions.

- Grain severely injured when the plant is 1 to 2 inches above the surface is reduced in yield by about 70 percent.

- Grain severely injured in the boot stage did not recover.

- Straws broken over just as the head is emerging from the boot exhibited a 50% yield reduction.

- Straws broken over as the head is nearing the ripening stage exhibited only 10% yield reduction. The author of this study concluded that hail adjustors must take into account the stage of growth at the time of adjustment when dealing with straws broken over. The exception to this is for straws broken such that the head hangs down to the level where the combine (the binder machine in 1937!) will not catch the head of grain.

A 1955 study at Oklahoma State University expanded on the 1937 work in Iowa and concluded that hail damage at the bloom and milk stages caused yield reductions not so much from mechanical damage from hailstones, but from reduced kernel size and loss of weight per 1000 kernels. In general, the greater losses were in the earlier stages, and as the wheat reached maturity the losses were less. [15]

The hail insurance industry has incorporated much of this early research into its indemnity calculations.

A 1986-1992 study in Spain entailed researchers placing a network of 250 hail collection pads in a grid formation over an area of 1000 km² situated inside a geographic target area in northern Spain. Over the seven summers of this study, 122 hail days were experienced. Barley and wheat crop damage in the test area was closely monitored and quantified. One key takeaway from this work was that wheat is more resistant to hail than barley. [16]

Cereal crops can be damaged by non-hail factors including: [17]

- Stem rust: this is caused by the fungus *Puccinia graminis*. It is evidenced by elliptical-shaped, powdery-red pustules on leaves and stems. Later in the season, their color will darken, hence the common name of black rust which is applied to stem rust.

- Leaf rust: this is caused by the fungus *Puccinia triticina*. It is otherwise known as brown rust owing to its reddish-brown lesions on leaves.

- Stripe rust: this is caused by the fungus *Puccinia striiformis*. It is otherwise known as yellow rust owing to its yellowish lesions on leaves.

- Fusarium head blight: this is caused by *Fusarium graminearum*. Symptoms will be discoloration on the glume tissues. As an aside, this blight poses a threat to the barley malting industry. The *Graminearum* fungus can lead to creation of *deoxynevalinol* which can affect the alcohol fermentation cycle and also lead to beer which profusely gushes when the drinker opens the can.

- Common bunt: this is caused by the fungus *tilletia*. Spores will be evident on the heads of grain.

- Grasshoppers: can feed on the un-ripened stem material below the head, weakening the head and causing it to fall off the plant.

- Wheat stem sawflies: can eat the stem tissues and weaken the stem.

- Wheat stem maggots: can work their way into the stem and impede the nutrient supply to the head causing a whitish coloration to develop.

Soybean *(Glycine soja)*

As noted in an earlier chapter, soybeans are classed as determinate and indeterminate. The indeterminate class is typically what would be found in the northern US states and in select locations in Canada. The adjusting discussion that follows is focused on indeterminate soybeans.

A 2011 article from the University of Nebraska provides excellent insight into how indeterminate soybeans are adjusted.[18]

Vegetative Stage Hail Storms

An early-stage hail storm can potentially destroy plants in the field. An adjustor will first determine the number of plants in a ten-foot section of row that has not had any stand reduction. Next, the adjustor will determine the number of plants in a typical ten-foot section of row that has been hail damaged. These plant counts will be converted into plants per acre using the formula:

(plants in 10 feet/row spacing) x 52,250.

The damage figures in Figure 7-4 indicate the yield loss damage as per the 2011 University of Nebraska study. For example, an original plant stand of 100,000 plants per acre damaged to leave only 50,000 plants per acre will have a stand reduction damage figure of 23%.

Original Stand	Remaining Stand											
	120	110	100	90	80	70	60	50	40	30	20	10
							Percent					
125	1	3	6	10	14	18	24	30	36	44	54	65
120	0	1	5	9	13	17	23	29	35	43	53	64
110		0	3	7	11	15	21	27	33	41	51	62
100			0	3	7	11	17	23	29	37	45	59
90				0	3	7	13	19	25	33	43	55
80					0	4	10	16	22	30	40	52
70						0	6	12	18	25	35	48
60							0	7	13	20	30	45
50								0	8	16	25	41
40									0	11	23	39

Figure 7-4

Leaf loss in the vegetative stage has little effect on the overall yield owing to the efficiency of the photosynthesis process. An adjustor will select a sample of 20 plants and make an estimation of the percentage of leaf area lost. Depending on the stage of growth at time of hail, and the leaf area destroyed, the chart in Figure 7-5 will provide a loss figure.

Growth Stage	Defoliation (% leaf area destroyed)									
	10	20	30	40	50	60	70	80	90	100
R1-2	0	2	3	5	6	7	9	12	16	23
R3	2	3	4	6	8	11	14	18	24	33
R4	3	5	7	9	12	16	22	30	39	56
R5	4	7	10	13	17	23	31	43	58	75
R6	1	6	9	11	14	18	23	31	41	53

Figure 7-5

Hail can break off plant branches or bend over plant branches all while leaving the plant surviving. The adjustor will examine the hail marks on the plants and determine the stage of growth at the time of hail. Next, a count will be taken of the number of nodes above the cotyledon node on 20 plants. The adjustor will count the number of nodes that are broken off. The data shown in Figure 7-6 shows the damage from broken nodes.

Growth Stage	Percent Nodes Cut Off						
	5	15	25	35	45	55	65
V1-Vⁿ	0	1	3	5	7	11	18
R1-R2	1	4	7	9	12	16	23
R2.5	2	6	10	14	18	24	32
R3	3	9	14	19	25	32	41
R3.5	4	12	19	27	35	43	53

Figure 7-6

This exercise will be repeated for stems (nodes) that are broken over but not broken off. The data shown in Figure 7-7 shows the damage from broken-over nodes. The defoliation damage (Figure 7-5) will be added to the damage figures from Figures 7-6 and 7-7 to arrive at an overall damage figure.

Growth Stage	Percent Nodes Cut Off						
	5	15	25	35	45	55	65
V1-Vⁿ	0	0	1	2	3	5	8
R1-R2	0	1	2	4	6	10	14
R2.5	1	3	6	9	11	16	20
R3	2	6	10	14	17	21	25
R3.5	2	8	13	18	23	28	33

Figure 7-7

Reproductive Stage Hail Damage

Late season hail injury that occurs directly to soybean pods will result in irreparable damage. Insurance companies will use a variety of techniques to asses pod loss damage. This could entail selecting a population of plants and counting pods on and off the plants. This could mean measuring off an area of row and counting the number of pods off the plants. One technique detailed by the University of Iowa equates 4 beans per square foot on the ground as being equal to a 1 bushel per acre loss. (19)

Canary Seed *(Phalaris canariensis)*

Canary seed is a difficult crop to adjust for hail damage. The individual plant stems are more resilient than other cereal grains. The small seeds are good and thoroughly packed inside the heads which offers the seeds protection from hail.

Adjusting canary seed will be similar to other cereal grains in that the hail adjustor will examine a population of plants for evidence of breakage due to hail. Next the heads will have to be examined for evidence that hail has dislodged seeds from the heads. As one looks down at the top of a head, if seeds are visible in the head, then that head has survived the hail storm. In my experience adjusting canary seed, I have yet to see much damage greater than 10% payable loss.

It is important to note that canary seed stems have a tendency to arc over as a result of the wind and rain that accompany a hail storm. The farm operator might argue that the hail storm has caused the stems to arc over and that he will have trouble picking up the plants with his combine header. While the arced-over stems might be difficult to combine, it is not hail that that has caused the stems to arc.

Faba Bean (Vicia faba)

Faba beans are not a widely grown in the US as these beans do not fare well in hotter, drier climate zones. Manitoba, parts of Saskatchewan and parts of Alberta do grow Faba beans.

Because Faba is not a widely grown crop, the hail insurance industry does not have a great deal of adjusting experience with the crop.

Faba will likely be treated in a manner similar to soybeans. An adjustor will examine the field for evidence of stand reduction in the case of an early-season hail event. For a hail storm in the vegetative growth stages, defoliation leaf loss will be assessed and a loss figure arrived at. While the procedure may not be identical to that of soybeans, it will be similar. In the case of a later stage hail event, a population of plants will be examined to determine pod loss or pod damage extent.

Buckwheat (Fagopyrum sagittatum)

The USDA[20] describes a general approach to hail adjustment on buckwheat. For hail events during stages N1 to N3 (flowering begins) the adjustment focus will be on stand reduction (plants destroyed). From stage N4 to N8, stand reduction and cutoff nodes/broken over plants are assessed for loss. From stage N9 and up, cutoffs and breakover will be the focus.

STAGE OF GROWTH	PERCENTAGE OF PLANTS DESTROYED																			
	5	10	15	20	25	30	35	40	45	50	55	60	65	70	75	80	85	90	95	100
N-1	0.0	0.0	0.0	0.0	0.0	0.0	0.0	0.0	0.0	0.0	0.0	0.0	0.0	3.5	14.5	26.5	40.0	55.0	71.5	100.0
N-2	0.0	0.0	0.0	0.0	0.0	0.0	0.0	0.0	0.0	0.0	0.0	0.5	3.0	8.0	18.5	30.0	43.5	58.0	74.0	100.0
N-3	0.0	0.0	0.0	0.0	0.0	0.0	0.0	0.0	0.0	0.0	0.0	1.5	6.0	12.5	23.0	34.0	46.5	60.5	76.0	100.0
N-4	0.0	0.0	0.0	0.0	0.0	0.0	0.0	0.0	0.0	0.0	0.0	2.0	9.0	17.0	27.0	37.5	50.0	63.5	78.5	100.0
N-5	0.0	1.0	2.0	3.0	3.5	4.5	6.0	7.0	8.0	9.5	10.5	13.5	20.0	27.5	36.5	46.0	57.0	69.0	82.0	100.0
N-6	0.5	2.0	3.5	5.5	7.5	9.5	11.5	14.0	16.0	18.5	21.5	25.0	31.5	38.0	46.0	54.5	64.0	74.0	85.0	100.0
N-7	0.5	3.0	5.5	8.5	11.0	14.0	17.5	20.5	24.0	28.0	32.0	36.0	42.5	48.5	55.5	63.0	71.0	79.5	88.5	100.0
N-8	0.5	4.0	7.0	11.0	14.5	18.5	23.0	27.5	32.0	37.0	42.5	47.5	53.5	59.0	65.0	71.5	78.0	84.5	91.5	100.0
	PERCENT OF LOSS FROM STAND REDUCTION																			

Figure 7-8

STAGE OF GROWTH	PERCENTAGE OF NODES CUTOFF/BREAKOVER																			
	5	10	15	20	25	30	35	40	45	50	55	60	65	70	75	80	85	90	95	100
N-4	0.0	0.0	0.0	1.0	2.0	3.0	5.0	7.0	9.0	11.0	13.5	16.5	20.0	24.5	29.5	35.0	41.0	47.5	55.0	62.5
N-5	0.0	0.0	0.5	2.0	3.5	5.0	7.0	9.0	11.5	14.0	17.0	20.5	24.5	29.0	34.0	40.0	46.0	52.5	60.0	67.0
N-6	0.0	0.0	1.0	2.5	4.5	6.5	9.0	11.0	14.0	17.0	20.5	24.5	28.5	33.5	39.0	44.5	51.0	57.5	64.5	72.0
N-7	0.0	0.0	1.5	3.5	6.0	8.5	10.5	13.0	16.0	19.5	23.5	28.0	33.0	38.0	43.5	49.5	55.5	62.0	69.5	76.5
N-8	0.0	0.0	2.0	4.0	7.0	10.0	12.5	15.0	18.5	22.5	27.0	32.0	37.0	42.5	48.0	54.0	60.5	67.0	74.0	81.0
N-9	2.0	3.5	6.0	8.5	11.5	15.0	18.5	22.0	26.0	30.0	35.0	40.0	45.5	51.0	57.0	63.0	69.5	76.0	83.0	90.5
N-10	3.5	6.5	9.5	12.5	16.0	20.0	24.0	28.5	33.0	37.5	42.5	48.0	53.5	59.5	65.5	71.5	78.0	85.0	92.0	99.5
N-11	5.0	7.5	10.5	14.5	20.0	25.5	31.0	36.5	42.0	47.5	53.0	58.5	64.0	69.5	75.0	80.5	85.5	91.5	96.0	100.0
N-12 and up	6.0	8.0	11.0	16.5	24.0	31.0	38.0	44.5	51.0	57.0	63.0	69.0	74.0	79.5	84.0	89.0	93.0	97.5	100.0	100.0
	PERCENT OF LOSS FROM PLANT DAMAGE																			

Figure 7-9

For a hail event at the combine-ready stage, damage assessment will entail a determination of seeds lost on the ground versus seeds remaining on the plant.

Sunflower *(Helianthus annuus)*

Hail damage to sunflowers can range from complete head obliteration to minor stem damage. Hail damage to these seemingly robust plants will be a function of hailstone size, speed, density, storm duration and plant stage of growth.

Hail damage in the early stages of vegetative growth (up to 40 days from planting) is not usually detrimental to crop yield as the damaged plants will likely recover.

Hail damage during stages R1 to R6 is a sensitive time for sunflowers. Defoliation of leaf matter can lead to a lack of photosynthesis and poor head development.

Hail damage near or right after the flower stage may not be detrimental to the longevity of the plant, but seed formation can be stunted. When researching this book, I did not come across any published hail adjustment studies or procedures.

Sugar Beet *(Beta vulgaris subsp. altissima)*

Sugar beet has effectively two parts to its growth cycle: vegetative growth when leaf material is growing, and sugar accumulation in the beet root.

Sugar beets in the vegetative growth stage have the ability to recover in the 30 or so days after a hail event. A 1955 study at Cambridge University in the U.K. revealed that little loss of crop occurred until 50% of leaf area had been defoliated or 50% of the plant density had been eliminated.

A 1994-95 study in Spain further supported the 1955 Cambridge study. Sugar beet plants at 7 different stages of growth were blasted two times, three times, and four times with a water jet to simulate light, moderate, and heavy hail damage. Results showed that the sugar content of the harvested beets deteriorated the most on the beets damaged in the latter stages of vegetative growth.

Later in the growing season, the plant reduces its top growth and starts accumulating sugar in the beet root. Defoliation from hail later in the growing season (September) will have little effect on crop yield. [21][22][23]

To Sum It Up

The hail insurance industry will offer the farm operator a payable loss amount based on the stage of plant growth and the severity of plant damage due to hail. These loss figures are not randomly arrived at. Over the past 80+ years, there have been numerous academic studies done in which hail damage has been simulated on plants. The ability of damaged plants to recover has been carefully observed. These academic studies form the basis for insurance industry rewards.

CHAPTER 8
MICRONUTRIENTS AND HAIL DAMAGE

In 2020, as part of my training, my instructor and I visited a canola field in southern Saskatchewan. We were taken aback at the severity of the damage to the stalks and branches of the plants. But we were also puzzled at how well the plants seemed to be recovering. When we eventually met up with the farm operator to review our damage findings, he told us that the day after the hail storm he had mobilized his hired men and had them spray micronutrients on all his hail-damaged crops.

The subject of micronutrients was again brought to my attention in late October, 2021 when I sat in on a presentation at Hick Seeds Ltd in Mossbank, Saskatchewan. The keynote speaker was from the Nachurs Alpine plant in Belle Plaine, Saskatchewan. He showed comparative images of various crops that had been treated with micronutrients and crops that were untreated. He mentioned that 2021 saw increased sales of micronutrients to areas of Saskatchewan affected by hail storms.

Following his presentation, I decided to include a chapter in this book discussing the role of micronutrients in assisting a damaged plant to recover from a stress event such as hail. Over the past twenty years, a significant amount of research has been done at the academic level to better understand the role of micronutrients in plant health and growth. In a search of the library database at Heriot Watt University for journal articles discussing micronutrient research, I was overwhelmed at the number of studies that have been done on crops ranging from mangos to onions, corn, wheat, lentils and even bananas.

A living plant is a system. If one feeds micronutrients into the plant system by spraying the nutrients onto the plant leaves (foliar application), the leaves will adsorb the nutrients. If the plant encounters physical injury from a stress event such as hail and if it realizes it has a deficiency in one or more nutrients, it will assimilate (absorb) the needed nutrients through its leaf structure to help repair itself.

Macronutrients and Micronutrients

A plant needs 17 nutrients if it is to exhibit maximum vegetative and reproductive growth. These nutrients are broadly classed as macronutrients and micronutrients.

Macronutrients include: Carbon, Hydrogen, Oxygen, Nitrogen,

Phosphorous, Potassium, Sulphur, Calcium, and Magnesium.

Micronutrients include: Iron, Zinc, Copper, Manganese, Molybdenum, Chlorine, Boron, and Nickel.

A review of a number of journal articles revealed the following broad tenets:

- Iron (Fe) can assist a plant by bolstering the photosynthesis process. Iron in combination with Molybdenum can help a pulse plant fix Nitrogen in the soil.

- Zinc (Zn) is a valuable cofactor in helping protein enzymes to optimally function.

- Copper (Cu) can bolster the photosynthesis process, increase pollen production and assist with plant cellular enzyme activity.

- Manganese (Mn) can help the chloroplast and aid the efficiency of the photosynthesis process. Manganese can also raise the plant's disease tolerance.

- Molybdenum (Mo) can assist with pulse crop Nitrogen fixation and can improve cellular enzyme activity.

- Chlorine (Cl) can help regulate the stomatal tissue to help the plant assimilate CO_2 which leads to improved photosynthesis.

- Boron (B) is a valuable constituent in enzyme activity.

- Nickel (Ni) is a cofactor in enzyme activity and also helps to convert urea into ammonia (NH_4^+). [1]

Some key research study findings that caught my attention as I read through a number of journal articles included:

- A 2017 joint study by Chinese and Pakistani crop scientists on three wheat varietals treated with a micronutrient mixture containing 117.5 mg/L Zn, 50 mg/L Fe, 7.5 mg/L Cu, 25 mg/L B, and 50 mg/L Mn saw the plants produce 3% more tillers, 17% more chlorophyll, and heads that were 36% longer. [2]

- A 2020 study in Nebraska showed that a micronutrient treatment containing Iron when applied at 123 grams per hectare to a corn crop displaying visual signs of nutrient deficiency raised harvested yield by 14 bushels per acre versus a similar distressed test patch left untreated. Results of this study also confirmed that the corn plant did take up micronutrients into the leaf tissue, although the plant did not use all the nutrients taken into the leaf. This study shows that a plant is a metabolic system that will use just the particular foliar nutrients the plant needs. [3]

- A 2015 study in China applied a micronutrient foliar treatment comprising 3 mls/L zinc sulfate to wheat at the milk stage. Compared to a non-treated test plot, there was no discernable benefit from the foliar treatment. Further evidence perhaps that if a plant does not need the nutrients, it will not use them. [4]

- A 2020 study in Poland using two potato varietals treated with 12 g/hectare Zn, 200 g/Hectare Mn, and 500 g/hectare B, displayed mixed results. This is further evidence that a plant will take up and utilize the nutrients if it needs them. [5]

- A 2020 study in India showed that a foliar application containing 0.5% Zn, 0.5% Fe, and 0.2% B to red lentils helped alleviate heat stress on the plants resulting in a 33% yield increase. [6]

- A 2018-2019 study in Iran on wheat treated with a foliar application containing 134 mg/L Fe and 270 mg/L Zn saw improved yield, increased kernel weight, and an increased flag leaf area (improved photosynthesis). [7]

The Future of Micronutrient Delivery

The mode of foliar application is changing rapidly. Dr. Mark Knell at the Nordic Institute for Studies in Innovation, Research and Education in Norway says that physics, chemistry, and biology are converging to create a technological revolution called *nanotechnology*. A nanoparticle by definition is a particle with one of its dimensions measuring between 1 and 100 nanometers (nm), with high surface tension and a high surface to volume ratio. [8]

Knell's argument is that by encapsulating the micronutrient substance in a nanoparticle, the particle will stick better to plant leaves (high surface tension), and the micronutrients will stand a better chance of being adsorbed into the leaf structure. The benefit of using this technology is both economic and environmental in that more foliar spray sticking onto the plant means less runoff during rain showers and therefore more efficient application.

CHAPTER 9
HAIL DAMAGE AND FUNGICIDES

As the global climate changes, temperature and moisture changes are creating an inviting set of conditions for the growth of fungal matter on agricultural crops. A 2012 study at Imperial College (London, UK) estimated that fungal infections annually destroy at least 125 million tonnes of the top five food crops: rice, wheat, maize, potatoes and soybeans. No doubt over the past decade, this figure has grown in magnitude. In addition, trees lost or damaged by fungi each year fail to absorb an estimated 230-580 megatonnes of atmospheric CO_2. This could contribute to an increase in greenhouse gas concentration which in turn could exacerbate climate change. [1]

A hail storm event is a form of precipitation in itself. In addition, a hail event is often accompanied by additional rain. If the days following the hail event are warm, the conditions for fungal infection on plants are prime. Moreover, a plant that has been stressed by the battering effects of a hail storm will have reduced natural defenses against the onset of fungal infections such as blight, rust, mildew, and mold.

Natural Defenses

A healthy plant can detect the efforts of a fungal intruder trying to incorporate its way into the leaf structure. A healthy plant can release an arsenal of defenses to thwart the fungal intruder including various reactive oxygen species, jasmonic acid, salicylic acid, ethylene, and callose. These various substances will harm the structure of the invading fungus. As a particular example, callose (molecules of glucose linked together) is synthesized and deposited between the cell wall and the invading fungus. The callose prevents the fungal cell from penetrating the plant cell. [2] Healthy plants can also defend themselves against fungal enemies by a defense system called *RNA silencing*. Fungal cells produce double-stranded RNA or DNA during replication. Plants can recognize these foreign molecules and respond by digesting the genetic strands into useless fragments and halting the infection. In addition, a healthy plant might retain a memory of the digested RNA material which can be used to quickly respond to future fungal attack. Healthy plants are also capable of releasing protein enzymes that can destroy the cell walls of an invading fungal intruder. [3]

Fungal Growth

A stressed, unhealthy plant will have a reduced ability to express these natural defenses. A fungal intruder will then waste little time in invading the plant's leaf and stem structure. Despite these best natural defense efforts by plants, fungal intruders often manage to succeed in penetrating the plant structure. Fungal intruders are part of the eukaryotic class of cells. By comparison, the other major class of cells is the prokaryotes, otherwise known as bacteria.

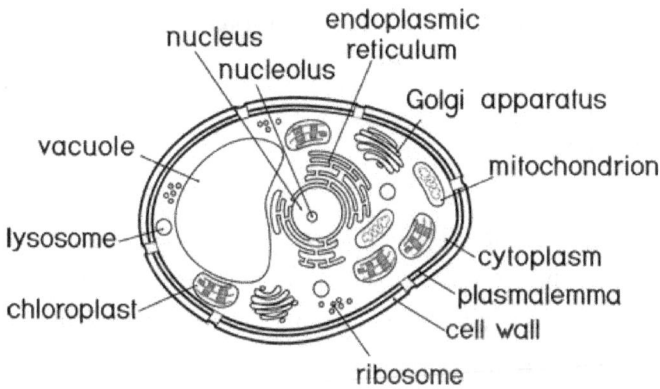

Figure 9-1
Eukaryote Structure

A eukaryotic cell differs from a prokaryote in its complexity. As Figure 9-1 shows, a eukaryote has a defined nucleus, a robust membrane structure containing peptidoglycan, and a cellular envelope around the various cell internal organelles.

Consumers who enjoy consuming beer and wine have plenty of experience with eukaryotic cells. Yeast is a eukaryote and is used to ferment malted grain mash in beer making and to ferment crushed grapes in winemaking.

A eukaryote is a heterotrophic organism in that it needs to consume food energy from living or dead organisms in order to survive. The cell wall structure of a eukaryote contains protein enzymes which are capable of degrading and then absorbing sugar-type molecules across the cell wall. Once inside the cell wall structure, the eukaryote will digest the sugar-type molecules that comprise plant tissue to produce the cellular energy needed to create new cells. This robbing of energy from the plant, quickly tilts the balance of power in favor of the eukaryote and the plant comes under assault by fungal growth.

Strobilurin

In the 1970s, a group of German researchers led by Dr. Timm Anke focused their attention on the creation of a molecule that could disrupt the cyclical process of eukaryotic growth. Their efforts led them to study an anti-fungal molecule on a pine cone fungus called *Strobiluris tenacellus*. This research direction was driven by prior work by scientists in Czechoslovakia who focused on the *Oudemansiella mucida* fungus from beechwood trees. The result of the work by Anke and his team was the isolation of *Strobulurin A* and *Strobulurin B* compounds. Figure 9-2 illustrates the structure of the strobulurin molecule. Anke and his team were able to show that when applied to the leaf structure of a plant, Strobilurin was able to disrupt the electron transport chain mechanism in the fungal mitochondrion. Without sufficient ATP energy, fungal cells cannot grow and proliferate.

In the 1980s, attention was turned to solving an issue unique to natural Strobilurin. The issue was the breaking down of the molecule when exposed to light. By appending various molecular structures to the natural Strobilurin molecule, scientists were able to achieve light stability. In 1992 Zeneca Agrochemical (now Syngenta) was the first to receive a patent on their synthetic Strobilurin-based fungicide called *Azoxystrobin*. BASF countered quickly with *Kresoxim-methyl Strobin*. Novartis (now Syngenta) joined the party later with its *Trifloxystrobin* product.

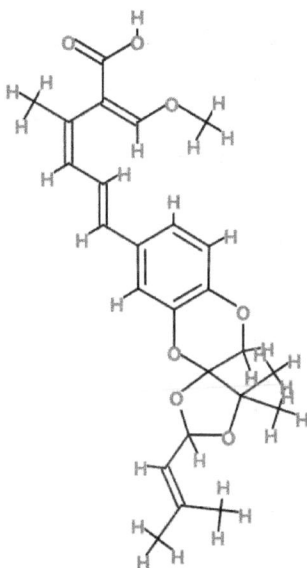

Figure 9-2
Strobilurin structure

Research has continued since the 1980s to identify new variants based on the Strobilurin backbone molecule. As a result of mergers and acquisitions since the 1980s, the global fungicide market is now dominated by only five companies: Syngenta (Switzerland), FMC Corp (USA), Bayer (Germany), BASF (Germany), and Adama (Israel).

Research efforts have identified new Strobilurin-based molecules including: triazole, chloronitrile, and dithiocarbamate. Triazole molecules when applied to a plant leaf structure will derail the ability of a fungal cell to produce sterol which is needed for growth. Chloronitrile molecules destroy enzymes in the cell wall of the fungal intruder and production of new fungal cells is stropped. Dithiocarbamate molecules act to bind up and deprive the fungal invader of Manganese and Boron atoms which it desperately needs to survive. Of these various compounds, fungicides based on Strobilurins A and B account for 25% of the fungicide market and are responsible for $2.5 billion in sales each year. [4]

CHAPTER 10
LAND LOCATIONS

A hail insurance adjustor's primary concern is finding his way to the correct hail-damaged field. In this tech-heavy age it is all too easy to just use a handheld GPS unit or a GPS smart phone app to navigate to a desired location to meet with the farm operator. An adjustor licensing exam will have a number of questions on it to test the applicant's understanding of land locations. This chapter is designed to help an exam candidate better appreciate land location systems. For readers curious about the history of land description systems, this chapter will also be of interest.

United States Land System

In the US, the land description system is called the *Public Land Survey System (PLSS)*. It is sometimes called the Rectangular Survey System.

This system was created by legislation called the Land Ordinance Act of 1785. The purpose of this system was to survey off parcels of the land that had been ceded to the United States by the Treaty of Paris in 1783 following the end of the American Revolution. This treaty awarded all lands east of the Mississippi River, north of Florida and south of Canada to the United States of America. Subsequent treaties from 1803 to 1848 saw the acquisition of all lands west to the Pacific Ocean.

Today the U.S. Bureau of Land Management (BLM) records the survey, sale, and settling of lands in the United States.

The PLSS is built around reference lines. There are six major north-south reference lines called Principal Meridians. Their locations are: 1st Meridian at 84 degrees 48 minutes longitude, 2nd at 86 degrees 27 minutes longitude, 3rd at 89 degrees 08 minutes longitude, 4th at 90 degrees 27 minutes longitude, 5th at 91 degrees 03 minutes longitude, and 6th at 97 degrees 22 minutes longitude. There are several smaller north-south lines scattered across various parts of states. For example, Black Hills Reference Line (South Dakota), Salt Lake Reference Line (Utah), Gila & Salt River (Arizona), New Mexico, Boise (Idaho), and Willamette Reference Line (Oregon-Washington) are some of the smaller north-south lines. There are a series of east-west baselines that complete the system. Some of these are: the 1855 Base Line along the Nebraska/Kansas border, the 1875 Base Line in Wyoming, the 1851 base line in California/ Nevada, the 1878 base line in South Dakota, and the 1870 Base Line in Oklahoma.

At a local level, land is divided into townships and ranges. A township comprises 36 sections of land, with each section measuring 1 mile x 1 mile. Within a given state, townships are referenced relative to whether they are east or west of the north-south meridian line. Townships are further referenced relative to whether they are north or south of a baseline.

As an example, North Dakota references the 5th Principal Meridian. A land location in North Dakota might read something like S7, T146N, R77W. This equates to section 7 in Township 146 North of the baseline, Range 77 West of the 5th Principal meridian.

Canadian Land Systems

In Canada the land registry system is more complex and involves four systems: *the Patchwork System*, the *River Lot System*, the *Rectangular System*, and the *Dominion Land System*.

The Patchwork System is still found in parts of Nova Scotia and Newfoundland. Land locations are referenced to a tree, a large rock or a river.

The River Lot system is found in parts of Quebec along the St. Lawrence River. Parcels of land that were about 10 x 15 kilometers in size were surveyed off perpendicular to the river. These parcels were then subdivided into smaller lots.

As land was surveyed off in Ontario, the Rectangular system came into fashion. Townships of size approximately 10 miles x 10 miles were identified. These parcels were then sub-divided into 100 or 200 acre lots for settlers. As a generic example, a land location might be identified as being Lot "D", Concession "A", Township "XYZ". On a map, the lots will appear as east-west parcels and the concessions will be rows of lots trending north-south. Individual townships are divided by east-west trending roads called *Side Roads* and north-south trending roads called *Lines*. At one time, to get to a particular location, one would have been told to travel along Side Road 10 until you reach the 5th Line. Head north on the 5th Line until you see the red barn and the white farmhouse. Today in Ontario, roads are individually named and farms identified with street/road numbers.

By the time western Canada started to open up to European settlers, surveyors realized a better system was needed. In 1871, the Dominion Land System was created. Its structure (based on Meridians) was influenced by the PLSS already in place in the United States.

- The First Meridian is located at 97 degrees, 27 minutes west longitude (just west of Winnipeg, Manitoba). The original plan was to start the first Meridian at Fort Garry (Winnipeg). This slight westerly shift was necessitated by the troubles at Fort Garry in 1871 involving personalities such as Louis Riel.

- The Second Meridian is at 102 degrees west longitude and forms the northern part of the Manitoba–Saskatchewan boundary.

- The Third Meridian is at 106 degrees west longitude and runs through Moose Jaw and Prince Albert, Saskatchewan.

- The Fourth Meridian is at 110 degrees west longitude and forms the Saskatchewan–Alberta boundary (bisecting the town of Lloydminster).

- The Fifth Meridian at 114 degrees west longitude runs through Calgary. (Barlow Trail in Calgary is built mostly on the meridian).

- The Sixth Meridian at 118 degrees west longitude runs near Grande Prairie, Alberta, and Revelstoke, British Columbia.

- The Seventh Meridian at 122 degrees west longitude runs between Hope and Chilliwack, British Columbia.

The east–west baseline used in surveying was taken as 49 degrees north latitude, which forms much of the Canada–United States border in the western provinces.

Starting at the 49 degree latitude baseline, surveyors marked off east-west numbering lines at 6 mile intervals. These intervals were called *Townships* (abbreviated Twp). Township 1 sits on the east-west baseline at 49 degrees latitude (the Canada-US border). Township 2 sits atop Township 1 and so on. This numbering progression continues northerly. (For example, in Saskatchewan by the time one reaches the city of Prince Albert, the township numbering will be at about Township 48. I live in Mossbank, Saskatchewan, at about Township 11).

In a westerly direction, surveyors started at a Principal Meridian and denoted intervals of 6 miles. These intervals were denoted *Ranges*.

A parcel of land created by moving north by 1 Township and west by 1 Range is thus 6 miles x 6 miles (36 sq. miles). This parcel is also denoted a *Township* and is written out in long form as *township*. This numbering nomenclature starts again at Range 1 once the next Meridian is reached. For example, in Saskatchewan when one reaches the town of Carnduff, the range is about 33. Several miles west at the town of Glen Ewan, the 2nd Meridian is reached and the numbering starts at Range 1.

Because the Earth is a sphere and not a square, some distortion will enter the picture. Townships will not all be 6 miles x 6 miles in size. To compensate, starting two Township lines north of 49 degrees latitude, a *correction line* was created. At each correction line the range is displaced a small distance (225 feet) west. Every 4th township line thereafter will have a correction line. Once a new Meridian is reached, the correction lines start again.

Access Permission

Once a hail adjustor reaches the land that will require examination and insurance adjusting, it should NEVER be assumed that the adjustor has the right to access the land on any vehicle with a motor and wheels. The insurance industry expectation and the farm operator's expectation are that the adjustor will walk the land to examine the crop. Insurance adjustors gets terminated each season because they failed to respect the land and the crop by driving into the field.

An exception to this is if the farm operator provides an ATV (all-terrain vehicle/quad) for accessing the field. In several cases, I have even had farm operators drive me on their ATV to the parts of the field where I needed to make my observations and counts. With late-season hail storms where combining is in progress, I will ask the farm operator if I can drive on his land with my truck to speed up the adjusting process. I have never had a refusal to my question of truck access in these situations. But I do exercise common sense. If the weather conditions are wet, I will never ask for truck access as that runs the risk of creating ruts in the field.

Frozen Fury

CHAPTER 11
EVIDENCE FOR THE ADJUSTOR

A hail insurance adjustor needs evidence to make determinations of damage. While this is not an issue when there are standing crops in the field, evidence is an issue if farm operators wish to harvest a hail-damaged field prior to the adjustor arriving on site.

Insurance companies will seek to have the adjustor obtain evidence from multiple portions of a field. Exact requirements will vary among insurance companies, but one common approach is to have the adjustor make one crop damage study per each 40 acres of land.

In a 160-acre field, the farm operator will leave evidence of damaged crop in each 40-acre portion of the field. In the case of standing crop, the combine operator should leave strips at least 20 feet wide and at least 40 feet long. The strips should be situated well into the field and away from the perimeter *headlands*.

In the case of a crop that has been windrowed (swathed), the combine operator should leave a piece of swath a good 40 feet long at four parts of a 160-acre field. The swaths should be situated well into the field and away from the perimeter headlands.

Evidence left for the adjustor should be untouched by the combine. Lifting the header and plowing ahead in the field to create a patch of standing crop evidence is unacceptable. Swerving the combine for a couple seconds to leave a patch 2 feet wide and 4 feet long is unacceptable. Advising the adjustor that the combine operator forgot to leave strips but that there are bits of standing crop at the very edges of the field is likewise unacceptable. The adjustor is within his rights to deny examining a crop where unacceptable or insufficient evidence is left.

CHAPTER 12
TECHNOLOGY AND THE FUTURE

The hail insurance industry is slowly embracing technology. Some companies now provide their adjustors with computer tablets to streamline the adjusting process. The tablet allows the adjustor to easily navigate to the affected crop using a built-in GPS function. Once on-site, the adjustor can enter his growth staging observations and damage count data into the tablet. The tablet will reference its programmed crop specific indemnity tables to calculate the percent damage losses owing to the farm operator.

We live in a society where technology is all too often taken for granted. People often do not stop to learn how technology works. A case in point is the GPS navigation system.

The GPS system is owned and maintained by the US government, Department of Defense. The GPS system comprises a total of 31 satellites in medium Earth orbit. This orbital level is defined as being 20,200 meters (12,550 miles) above the surface of the Earth. The satellites are tilted at an angle of 55 degrees to their orbital planes. Each satellite circles the Earth twice per day. The US Government has committed to having a minimum of 24 of the satellites active 95% of the time. The orbital space around Earth has been divided into six orbital planes. Each plane will thus have a minimum of four operating satellites 95% of the time. This geometric arrangement ensures that a person anywhere on Earth will be able to receive a signal from at least four satellites at any time of day (subject to the 95% up-time factor). Figure 12-1 illustrates the orbital planes. Each satellite broadcasts three signals at frequencies of 1575.42 MHz, 1227.60 MHz and 1176.45 MHz. The 1227.60 MHz frequency is reserved for military use. The other two are available for civilian use.

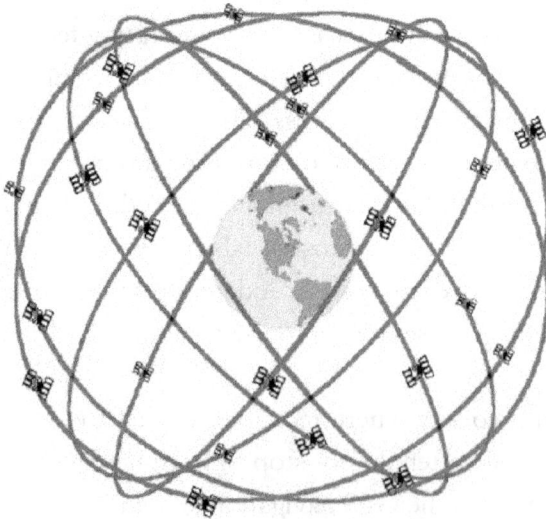

Figure 12-1
GPS orbital planes

Each satellite is equipped with an atomic clock that records extremely accurate time. A GPS receiver unit measures the time between signal broadcast and signal receipt. A minimum of three signals will allow the GPS receiver unit to calculate by triangulation the longitude, latitude and elevation. Less expensive receivers, such as those in smartphones, will utilize only one frequency (1575.42 MHz or 1176.45 MHz). The accuracy of a less expensive receiver unit will be to within 4.9 meters (16 feet). More expensive receiver units (such as those used in a tablet carried by a hail adjustor) will use both civilian satellite signal bands. The accuracy of location will be within 0.643 meters (2.1 feet).

The US Government is not alone in using satellites for positioning. The Russian Federation operates a network of 24 satellites at 19,100 meters elevation (GLONASS system). The European Union has 14 satellites in orbit (Galileo system). The Chinese have 22 satellites in orbit (Bei Dou system), and India has 7 units in orbit (NAVIC system). [1][2]

The current model revolves around a hail adjustor making a physical visitation to an affected field. Visual observations and physical counts form the basis for the damage level calculated by the insurance indemnity tables. But, drone technology could one day soon reduce the amount of physical time an adjustor spends in a field.

A drone is defined by Canadian and US transportation authorities as a remotely piloted aircraft system (RPAs). The legal expression unmanned aerial vehicle (UAV) is also applied. The prevailing aviation regulations in Canada and the US extend to drones. In December 2021, I wrote my RPAs exam. Studying for it took me back into the textbooks I used decades ago for my pilot's license. The exam was brutally hard, as it should have been. Transportation authorities do not want thousands of hobbyists messing around with drones. A difficult exam means only serious players will be allowed to operate drones. I passed the exam and plan to purchase a drone early in 2022.

Academic literature suggests that sensor data captured by a drone flying over a crop field can reveal significant information about the field. The data capture by drones is related to the physics of wave energy. Light energy from the Sun hitting an object is partly absorbed and partly reflected. A sensor on a drone can capture this reflected wave energy. Consider the energy spectrum in the table shown in Figure 12-2: [3]

Wavelength (meters)	Type of Wave
1×10^{-14}	Gamma
1×10^{-12} to 1×10^{-10}	X-Rays
1×10^{-8} to 1×10^{-7}	U-V Rays
1×10^{-7}	Visible
1×10^{-6} to 1×10^{-4}	Infrared & NIR
1×10^{-2}	Microwave
1	FM radio
100	TV
10,000	AM radio

Figure 12-2
Wavelength Spectrum

The data in Figure 12-2 is arranged with shorter wavelengths at the top and longer wavelengths at the bottom. A shorter wavelength has a deeper penetration.

- When we fly on an airplane at near 40,000 feet we are exposed to harmful Gamma rays in the atmosphere. The metal body of the aircraft absorbs much of this Gamma radiation, but some of it penetrates the aircraft structure and is absorbed by our bodies.

- When we visit a hospital with a broken bone we are often given an x-ray which penetrates our body and creates a photographic image of our broken bone.

- When we go outside without sunscreen, we stand the risk of getting a sunburn from U-V rays coming from the Sun.

- The microwave oven we use in our kitchen to warm food emits energy of wavelength 0.01 meters (1×10^{-2} m).

- Our favorite FM and AM radio stations operate on longer wavelengths of up to 10,000 meters.

Color Wavelengths

In the middle of the light energy spectrum is a category called Visible. The human eye is structured to accept wavelengths from 3.8×10^{-7} meters to 7.5×10^{-7} meters. The optic nerve in the eye impacted by this energy transmits a signal to the brain which causes us to interpret the wavelength energy as color.

What we interpret as purple color has a wavelength of between 3.8×10^{-7} meters to 4.5×10^{-7} meters. Blue is interpreted from wavelengths around 5.6×10^{-7} meters. Green is interpreted from energy of wavelength 4.8 to 5.6×10^{-7} meters. Yellow is interpreted from energy of wavelength 5.6 to 5.7×10^{-7} meters. Orange is interpreted from energy of wavelength 5.8 to 6.2×10^{-7} meters. Red is interpreted from energy of wavelength about 6.2 to 8×10^{-7} meters.

Taking canola as a practical example, when light hits a canola plant during early vegetative growth, the plant reflects some of the light energy in the 4.8 to 5.6×10^{-7} meter range. The human eye interprets this energy as the color green. As the canola starts to flower, the flower tissue reflects light in the 5.6 to 5.7×10^{-7} meter range. The human eye interprets this energy as the color yellow.

Recording color is not limited to just the human eye. Cameras also record light energy. There was a time when we would load our camera with a roll of film. The film was a piece of plastic impregnated with particles of silver chloride, silver iodide and silver bromide (silver halides). Opening the shutter on the camera exposed a section of the film to the light. The various wavelengths of light caused the silver particles to alter their shape. When the film was sent away to a film lab for processing, the film was exposed to various chemicals which caused the silver halides to convert to silver, creating a negative which in turn was used to print photos with a further chemical process. [4]

Technology has largely replaced rolls of film. Cameras today are digital, equipped with a data storage chip instead of a roll of plastic film. Light entering the camera is passed through a color array filter which allows red, green, and blue wavelengths to pass through. These wavelengths are referred to as RGB light. The data capture chip records the RGB energy across an array of pixels. An algorithm in the camera examines each pixel and interpolates the RGB data into the array of colors that we immediately see when we examine the picture image. The raw data for the pixel array is stored in the camera in uncompressed format as a TIFF file. When we download the TIFF data onto our computers, software compresses the data into a JPEG file format. [5]

Cameras are no longer hand-held devices. Cameras can be mounted on a drone and a drone operator can obtain RGB images as the drone unit flies above a crop field. The uncompressed TIFF file data can be processed through software. In 2016, researchers in the south of France used a 24 mega-pixel RGB camera of 60 mm focal length mounted to a drone. The camera was tilted at an angle of 45 degrees. The camera was turned 90 degrees to the direction of flight. The drone was flown 3 to 7 meters above ground and parallel to the row direction in a wheat field. Data was analyzed using *Agisoft* software. The stage of growth of the wheat crop was pre-flag leaf. Data analysis and calculation of plant density was within 10% of the actual seeding density. Simple RGB imagery has the potential for significant application in agricultural crop analysis. In the case of a field exhibiting stand reduction due to hail damage, RGB image data should be able to provide a close estimate of damage levels. [6]

Camera sensing technology has expanded to include the entire array of visible light wavelengths. Drones can be equipped with multispectral cameras. A 2017 study at Nanjing University in China used a drone to capture imagery of a wheat field at five stages of growth (jointing, booting, heading, germination, maturation). The multispectral camera attached to the drone collected data at 8×10^{-7}, 7×10^{-7}, and 5.50×10^{-7} meters wavelength. The data was used to calculate a vegetative index for each growth stage. The calculated indices aligned closely to ground-truthing observations. Multispectral imagery data holds potential for assessing damage levels of cereal crops that have been hit by hail. [7]

A 2016 study at Washington State University involved simulated hail damage to a potato crop. A drone was used to capture spectral data within 10 days of the damage. The camera attached to the drone was set to capture wavelengths of $4.8-5 \times 10^{-7}$, $4.5-4.8 \times 10^{-7}$, and $7-14 \times 10^{-7}$ meters. These wavelength bands are more commonly known as green (G), blue (B), and near infrared (NIR). The collected data was used to calculate the Green Normalized Vegetative Index (GNVDI) which is given by the formula:

$$GNDVI = (NIR - G)/(NIR+G)$$

Results of this study showed that this index correlated to the final potato yield with 77% accuracy for data captured within 10 days of damage. The authors of the study note that more work is required, but this study underscores the potential for multispectral imagery in assessing plant hail damage. [8]

In the coming years it will be interesting to observe how spectral imagery captured by drones finds its way into the hail insurance industry. With my drone and soon to be launched drone business I certainly will be watching for further academic studies that support the use of drones to assess crop damage. In the meantime, I will be working with farm operators in southern Saskatchewan to assess their crops at various stages of growth to delineate nutrient stress and disease stress, both of which will translate into differences in reflected RGB and NIR light wavelength energy.

Frozen Fury

FINAL REMARKS

This book has taken the reader on a journey through weather, climate change, hail models, seed structure, germination, crop growth, insurance compensation, land description systems, and technology.

For the farm operator, I hope this book has provided you with some new knowledge, or at the very least served to refresh your memory on plant details you may have forgotten.

For the trainee hail adjustor, I hope this book will be of value in helping you study for your licensing exam.

My brief tenure as a hail adjustor has been extremely rewarding. I have learned a considerable amount of new scientific information. I hope to continue doing some adjusting each summer. Hail adjusting has allowed me to see parts of my province that I would not otherwise have seen. Along the way, I have met some fascinating people. Moreover, I have gained a profound new appreciation for the risk-taking farm operators who grow our food.

My advice to anyone thinking about becoming a hail adjustor is – go for it.

GLOSSARY

Abiotic stress: a non-living factor that acts to stress living organisms. The stresses include drought, low or high temperatures, wind, and other environmental extremes including hail.

Adiabatic System: a thermodynamic system in which no heat energy is added.

Agrobacterium tumefaciens: the soil bacteria from which a genome was extracted and incorporated into soybeans to make them Round Up ready.

Alberta Hail Project: a long-running cloud modification program located in Alberta since 1996. This cloud seeding project uses the same contractors as the **NDCMP** program.

Anthropogenic climate change: climate change caused by human activity. This is how climate change is typically described by University-level educators.

Arctic Amplification: a model proposed by Harvard professor Lei Wang who argues that as Arctic ice quantities are reduced, the Arctic waters take on more solar heat energy. This energy serves to thicken the troposphere over the northern latitudes. This, in turn, affects circulations in the Hadley, Polar, and Ferrell cells.

Atmosphere: the mass of air that envelopes planet Earth. The atmosphere is comprised of several layers: the troposphere (closest to Earth), the stratosphere, the mesosphere, the thermosphere, the ionosphere, and finally space.

ATP: more formally known as Adenosine Tri-phosphate. ATP captures cellular energy and releases it as required to drive cellular reactions.

Beneficial Competition: assumes there is a deficiency of tiny nucleation sites in a cloud formation. It assumes the injection of silver iodide will produce a significant number of ice nucleation sites. The existing natural nuclei sites and the silver iodide ice crystals will compete for the available supercooled liquid cloud water within the cloud.

Bernard Vonnegut: 1940s scientist at General Electric credited with discovering that hail formation can be initiated by seeding a cloud formation with tiny silver iodide particles.

Calvin Cycle: as light energy from the Sun hits the chlorophyll, a residual phosphorous atom (Pi) present in the leaf tissue helps in the conversion of the photon light energy to ATP energy. The residual phosphorous atom comes from the Calvin Cycle which is the other half of the overall photosynthesis process. The Calvin Cycle comprises three stages: Carbon fixation, Reduction, and Regeneration.

Cellulose: in a plant or seed structure, the primary cell wall is built mostly of cellulose which is a chain-like structure of glucose $(C_6H_{12}O_6)$ molecules.

Chlorophyll: a green-colored pigment in plant cells which is capable of absorbing light.

Chloroplast: each mesophyll cell contains smaller structures called chloroplasts.

Coleoptile: in a monocot seed, the epicotyl tissue from which leaves form is sheathed in this tissue.

Coleorhiza: the sheath tissue that protects the radicle root.

Collar method: just as the collar of your shirt wraps part way around your neck, a leaf will wrap part way around the corn stalk structure. It is only when a leaf exhibits this wrapping feature that it is assigned a stage count (V1 to Vn).

Cotyledon: see Scutellum.

Cryptochrome: a protein substance in plant tissue that responds to light of the blue wavelength.

Climate: the weather conditions over long periods of time.

Cold Front: a cold air mass bumping against a warm air mass will cause the warm air mass to rise. Moisture in the warm air mass will precipitate and form larger, more billowing clouds.

Correction Line: a road jog to correct for the northward convergence of meridian lines.

Dew Point: the temperature at which a mass of air attains full saturation.

Dicotyledons: one of two broad classifications of plants. Dicots have a thicker, denser root system built around a main tap root.

Disappearing Deductible: a hail insurance policy in which the farm operator will only receive a financial settlement if the adjustor assigns a damage amount equal to or greater than the stated disappearing deductible amount. The deductible reduces on a sliding scale as damage levels increase.

Dominion Land Registry System: once surveyors began allocating parcels of land to settlers heading west, it was realized the Rectangular System was not practical. In 1871 the Dominion Land System was created and was influenced by the PLSS already in place in the United States. The First Meridian is located at 97 degrees, 27 minutes west longitude (just west of Winnipeg, Manitoba).

El Nino: the weather pattern created when the Trade Winds slow. Weather patterns can turn wetter along the west coast of North America and drier in south-east Asia, India and Australia.

Epicotyl tissue: that which leads to formation of plant leaf material.

Epigeal germinator: plants in which the cotyledon tissue is pushed above the soil are said to exhibit epigeal germination.

Erucic Acid: a long-chain, mono-unsaturated, omega-9 fatty acid molecule not suited for human consumption. This acid is present in canola (rapeseed) in small amounts and in mustard in larger amounts.

Eukaryote: a cellular organism having a defined nucleus, a robust membrane structure containing peptidoglycan, and a cellular envelope around the various cell internal organelles (ie fungus).

Fall Speed: the terminal velocity of a falling hailstone, given by $Vt = \sqrt{(2mg)/\rho AC}$

Ferrell Cell: circulation that occurs between about 30 degrees and 60 degrees of latitude. Ferrell Cells link Polar Cell flow and Hadley Cell flow. Ferrell Cells create wind flow from east to west in the equatorial region.

Flavonoids: polyphenol ring structures attached to a carbon atom backbone found in most fruits and vegetables. Flavonoids have powerful anti-oxidant properties. Flavonoids also play a role in the Nitrogen fixation process.

Full Coverage: a style of policy which provides for a set amount of money per acre depending on the percent damage to the crop.

Gibberelic Acid: an acid generated in the embryo that causes endosperm proteins to modify.

Glucose: see Cellulose.

Glucosinolates: molecules in rapeseed which in the presence of water make isothiocyanates which interfere with iodine uptake via an animal's thyroid gland.

GPS (Global Positioning Satellite) System: the GPS system comprises a total of 31 satellites in medium Earth orbit. This orbital level is defined as being 20,200 meters (12,550 miles) above the surface of the Earth. Each satellite circles the Earth twice per day. The US Government has committed to having a minimum of 24 of the satellites active 95% of the time. Each satellite broadcasts three signals at frequencies of 1575.42 MHz, 1227.60 MHz and 1176.45 MHz.

Hadley Cell: warm air at the equator rises to a height of up to about 18 kilometers (11 miles) above Earth. As the warm air rises, it spreads northerly and southerly on either side of the equator. By the time the warm air has reached about 30 degrees of latitude on either side of the equator, large portions of the warm air cool, start to sink, and flow back to the equator region. This cellular circulation pattern is called a Hadley Cell.

HAILCAST: the HAILCAST model begins with a liquid droplet at the base of a storm cloud that has been predicted to occur by the NARCCAP model. The liquid embryo droplet (300 μm in size) is assumed to experience an updraft of 4 meters/second. When the droplet encounters the layer of the storm cloud that is minus 8°C, the droplet freezes. The frozen droplet is assumed to contain some air giving the droplet a density of 900 kg/m³. The droplet continues to grow so long as the temperature in the cloud is between minus 20°C and minus 40°C.

Headlands: the perimeter area around a field. During seeding, the farm operator may end up turning his equipment around on the headland area and thereby double-seeding. The headlands should be avoided when attempting to study the crop for hail damage.

Hemicellulose: in a plant structure, cell walls may also contain hemicellulose which is a chain-like structure comprised of 5-carbon variants of glucose. In the leaf cell structure, hemicellulose chains are cross-linked to the cellulose chains.

Heterosis: plant vigor, in the context of Hybrid canola.

Heterotrophic: a cellular organism that needs to consume food energy from other living or dead organisms in order to survive.

High Clouds: Cirrocumulus, Cirrus and Cirrostratus clouds comprise the high cloud category. These formations are described as 'wispy', 'feather-like', or 'cotton-like'.

Holocene Epoch: the post-Ice Age period following the Pleistocene Epoch.

Humidity: the moisture content of the air in the lower atmosphere.

Hypocotyl tissue: the tissue that leads to formation of plant stem material.

Hypogeal germinator: plants in which the cotyledon tissue remains beneath the soil are said to exhibit hypogeal germination.

Ice Crystal Region: the upper part of a cumulonimbus cloud formation.

Jet Stream(s): Rossby waves, along with the rotation of Earth, create movements of air in patterns called the jet streams. The jet streams can best be compared to loops of ribbon.

Jointing: stem elongation on a cereal grain plant.

La Nina: the weather pattern created when the Trade Winds speed up. Weather patterns can turn drier along the west coast of North America and wetter in south-east Asia, India and Australia.

Lapse rate: the rate at which a mass of air loses temperature with rising altitude. The lapse rate is generally taken to be a temperature loss of 1.98°C per 1000 feet of altitude gain.

Leaf Area Index: a measure of leaf area per unit of ground area around the plant. Used especially in canola and sugar beet crop evaluations.

Legume: a legume plant is capable of interacting with *Rhizobia* bacteria in the soil to fix Nitrogen into the sub-surface soil.

Long Day: long day plants will start to flower (reproductive growth) when darkness hours are at or near a minimum (around the Summer Solstice).

Low Clouds: Stratus, Stratocumulus, and Nimbostratus clouds comprise the low cloud category. These puffy clouds and usually between 600 to 900 meters (2000 to 3000 feet) above the ground. These formations are associated with the development of cold fronts.

Ludlum, F.H.: a renowned hail researcher at Imperial College in London, UK. Noted for his 1966 paper, Cumulus and Cumulonimbus Convection, where he described up-drafting convection and cloud formation.

Macronutrients: Carbon, Hydrogen, Oxygen, Nitrogen, Phosphorous, Potassium, Sulphur, Calcium, and Magnesium.

Micronutrients: Iron, Zinc, Copper, Manganese, Molybdenum, Chlorine, Boron, and Nickel.

Mesophyll Cells: cells in the middle of the leaf structure are called mesophyll cells. These cells are where photosynthesis occurs.

Microfibrils: fibers of glucose molecules which provide the mechanical strength to the cell walls of a plant leaf.

Middle Clouds: Altocumulus and Altostratus clouds comprise the middle cloud category. Altostratus clouds are also associated with warm fronts.

Monocotyledons: one of two broad classifications of plants. Monocotyledons (monocots) generally have a fibrous root system that branches off in many directions.

NADPH: properly known as Dihydronicotinamide-adenine dinucleotide phosphate, NADPH is the molecule NADP that has gained Hydrogen atom (and thus one electron). NADPH is fundamental in making cellular reactions run.

Nanotechnology: a nanoparticle is a particle with one of its dimensions measuring between 1 and 100 nanometers (nm), with high surface tension and a high surface to volume ratio.

NARCCAP: weather models in North America largely center around The North American Climate Change Assessment Program (NARCCAP). This is an atmospheric-ocean circulation model that can be run based on a number of scenarios.

NDCMP: North Dakota Cloud Modification Project, which has been in existence since 1951.

NOAA: National Oceanic and Atmospheric Administration.

Pectin: a carbohydrate structure (carbon, hydrogen, oxygen) that helps to bind together individual cells and regulate water uptake by the cells.

Photoperiodicity: plants responding to varying amounts of daylight is called photoperiodicity.

Photosynthesis: the chemical mechanism that drives plant growth. In the photosynthetic process, the plant leaf tissue takes in carbon dioxide (CO_2) and expels oxygen (O_2).

Phytochrome: a protein substance in plant tissue that responds to light of the red wavelength.

Pleistocene Epoch: the period of the last Ice Age.

Polar Cell: as warm air from the equator heads poleward, it cools and begins to sink and head back towards 60-70 degrees latitude. This thermal-polar circulation pattern is called a Polar Cell.

Pressure Gradient: warm air flowing from a warm column towards a cold column creates a Pressure Gradient.

Prophyll: on a cereal grain plant, the base of a tiller is protected by tissue called the prophyll.

Public Land Survey System: Public Land Survey System (PLSS) and is sometimes called the Rectangular Survey System. This system was created by legislation called the Land Ordinance Act of 1785. The purpose was to survey off the land that had been ceded to the United States by the Treaty of Paris in 1783, following the end of the American Revolution. This treaty awarded all lands east of the Mississippi River, north of Florida and south of Canada to the United States of America. Subsequent treaties from 1803 to 1848 saw the acquisition of all lands west to the Pacific Ocean.

Pulse Crop: a legume plant from which ripened seeds are harvested for human consumption.

Rectangular Land Survey System: as land was surveyed off in Ontario, Canada townships of size approximately 10 miles x 10 miles were identified. These parcels were then sub-divided into 100 or 200 acre lots for settlers. A land location is identified as being Lot "D", Concession "A", Township "XYZ", where the lots are divided into east-west parcels and the concessions are rows of lots trending north-south. Furthermore, individual townships are divided by east-west trending roads called Side Roads and north-south trending roads called Lines.

Reinsurance: a hail insurance company will be backed financially (insured) by a larger insurance company. Reinsurers generally offer two types of **reinsurance:** Treaty and Facultative. For hail insurers, the Treaty method will often apply in that the re-insurer agrees to back all the policies written in a given year for a specific type of insurance. Re-insurance may be considered proportional or non-proportional. Under the proportional model the re-insurance firm receives a percentage share of all policy premiums sold by the hail insurance company. If a hail claim is recorded by a farm operator, the re-insurance firm bears a portion of the losses based on a pre-negotiated percentage. With non-proportional reinsurance, the re-insurance firm is liable if the hail insurance company experiences losses that exceed a specified amount. This amount is known as the retention limit.

Relative Humidity: the measured amount of humidity present in the air expressed as a percentage of the amount of humidity that would be present if the air were fully saturated.

Rennick and Maxwell: Alberta-based climate scientists known for their nomograph model which characterizes the sizes of hailstones as ranging from shot size to larger than golf-ball size.

Reproductive Growth: Nature propagates itself by generating seeds and fruits in the reproductive growth phase after vegetative growth is complete.

Ridges: the top parts of the jet stream.

Rossby Waves: the motion of air in the Polar Cells occurs in multiple waves, called harmonic waves, or Rossby Waves.

Roundup Ready: plants that are Roundup Ready have been genetically engineering to include a gene segment from the soil bacteria *Agrobacterium tumefaciens* into the plant genome. This advancement allows a farm operator to top dress the crop to kill weeds while not harming the plant.

RPAs: remotely piloted aircraft, otherwise called a drone.

RuBisCo: in the photosynthetic process, the critical enzyme that makes the reaction run is ribulose-1,5-bisphosphate carboxylase-oxygenase. For simplicity, it is nicknamed RuBisCo.

Scutellum: in a monocot seed, embryo tissue is distinguished from the endosperm cell structure by this membrane of vascular tissue.

Seed Structure: the three main components of a seed are: seed coat, the endosperm, and the embryo (consisting of epicotyl, hypocotyl, radicle, and cotyledon tissues).

Short Day: short-day plants will start to flower when dark hours exceed a minimum threshold (late in August).

Stomata: the surface of a plant leaf contains pores called stomata which allow the leaf to take in CO_2 and expel O_2.

Storm Clouds: clouds of the cumulonimbus variety that are often the origin of hail events.

Straight Deductible: the payable losses arising from hail damage will be reduced by a deductible amount stated in the policy.

Strobilurin: a chemical structure of fungicide derived from a pine cone fungus called *Strobiluris tenacellus* and the *Oudemansiella mucida* fungus from beechwood trees.

Strong Jet Stream: a strong jet stream is comprised of winds that are intense and compressed.

Supercooled: large water droplets will freeze at a temperature at or just below 0°C (32°F). Smaller droplets (less than 1 mm in size) can stay in liquid form down to minus 40°C (minus 40°F). These droplets are said to be supercooled.

Thermals: updrafting air currents associated with storm clouds.

Thylakoids: inside each chloroplast are structures called thylakoids which contain a green-colored pigment (chlorophyll) which is capable of absorbing light.

Trade Winds: see Ferrell Cells.

Trichomes: tiny, hair-like fibers attached to the seed hulls of canary seed which render the seed unfit for human consumption.

Plant breeding efforts have now created varietals that do not express trichomes (glabrous canary seed).

Troposphere: the component layer of the atmosphere ranging in height from about 8500 meters (28,000 feet) at the North and South poles to about 16,000 meters (54,000 feet) at the equator.

Troughs: the lower parts of the jet stream are called troughs.

µm: this symbol denotes a micrometer which is 1×10^{-6} meters.

µmol: this symbol denotes a micro-mol, which is 1×10^{-6} of a mol. A mol is calculated as the mass of a substance divided by its molar mass from the Periodic Table.

Vegetative Growth: the stage of growth of a plant that displays emergence of tissue material from the seed, formation of stalks, stems, secondary branches and leaves.

Vertical Clouds: (also called Convection Clouds) Cumulus and Cumulonimbus clouds comprise this cloud category. A cold front formation is associated with this cloud category. Cumulus and Cumulonimbus clouds develop by accumulating moisture. Up-drafting air currents, called thermals, complete the process.

Warm Front: if a faster-moving warm air mass begins to interact with a slower-moving cold air mass, the warm air mass will rise over top of the cold mass.

WBF Model: more formally called the Wegener-Bergeron-Findreisen Model. The period 1910 through the late 1930s was a vibrant time for research into cloud formations. German scientist Alfred Wegener, Norwegian scientist Tor Bergeron and Czech scientist Walter Findreisen developed this model to describe hail formation.

Weak Jet Stream: a weak jet stream is comprised of weak winds.

Weather: the state of the atmosphere at a given moment in time in a particular geographic location.

Weather Disturbance: an interaction of air masses of different characteristics.

Wind: the movement of air within a pressure gradient.

Younger Dryas: the Holocene Epoch transition to warmer temperatures would encounter an obstacle that would present Neolithic man with a challenge. Starting about 13,000 years ago, and lasting for about 1300 years, the global climate shifted back to being cooler and drier.

Zadoks method: a system of describing cereal plant growth developed by Dutch crop scientist Jan Zadoks in the 1970s.

ABOUT THE AUTHOR

When people ask Malcolm Bucholtz what he does, he asks if they want the long answer or the short answer.

The short answer is, he does a lot of things.

The long answer is:

After graduating from Queen's University with a degree in Metallurgical Engineering, he spent 16 years in the Canadian steel industry in roles ranging from Research & Development to Operations Management.

After completing his MBA degree from Heriot Watt University in

Scotland he became an Investment Advisor and Commodity Trader with a small, independent brokerage firm in western Canada.

The lure of the mineral exploration industry proved too great and he eventually left the brokerage industry to become President/ CEO of a publicly traded mineral exploration company focused on rare earth mineral deposits in New Mexico.

After years of travelling back and forth to the US, he decided to slow down a bit and more fully pursue his hobby of brewing and distilling. But slowing down was apparently not to be. As the craft distilling phenomenon gripped Canadian entrepreneurs, he started teaching these entrepreneurs technical courses related to brewing and distilling science. This led to returning to Heriot Watt University in 2017 to pursue a M.Sc. degree in Brewing & Distilling Science. Since completing his M.Sc. degree he has authored two books on the subject of distilling. A book exploring the deep science of beer brewing is currently in the works for release later in 2022.

When not delivering technical courses on brewing or distilling (www.ProhibitionUniversity.com), he trades the financial markets using mathematical techniques rooted in astrology, astronomy, and Sun barycenter cycles. He writes a bi-weekly subscription-based financial astrology newsletter (www.InvestingSuccess.ca). Each year he publishes a Financial Astrology Almanac.

He is the author of nineteen books in total.

NOTES

Chapter 1

(1)   Prein, A., Holland, G. (2018) Global Estimates of Damaging Hail Hazard. *Weather and Climate Extremes*. 22, pp:10–23

(2)   Plummer, C., McGeary,D. (1980) *Physical Geology*. McGraw Hill, Canada.

(3)   Hays, J., Imbrie, J., Shackleton, N. (1977). Variations in the Earth's Orbit: Pacemaker of the Ice Ages. *Science*. pp:. 1121-32.

(4)   NASA Observatory website (2000) Milutin Milankovitch. [online] Available at: https://earthobservatory.nasa.gov/features/Milankovitch. Accessed: December 2021.

(5)   Yale Program on Climate Change (2021) International

Public Opinion on Climate Change. [online] Available at: https://climatecommunication.yale.edu/publications/ international-public-opinion-on-climate-change/3/. Accessed: November 2021.

(6)   Encyclpedia.com website (2018) Jet Stream. [online] Available at: https://www.encyclopedia.com/earth-and-environment/atmosphere-and-weather/atmospheric-and-space-sciences-atmosphere/jet-stream. Accessed: October, 2021.

(7)   BBC website (2021) Atmosphere and Climate. [online] Available at: https://www.bbc.co.uk/bitesize/guides/ zpykxsg/revision/1. Accessed: October 2021.

(8)   United Nations website (2021) Population. [online] Available at: https://www.un.org/en/global-issues/ population. Accessed: November 2021.

(9)   C2ES website (2021) Global Emissions. [online] Available at: https://www.c2es.org/content/international-emissions. Accessed: November 2021.

(10)   Voosen, P. (2020) Why does the weather stall? New theories explain enigmatic 'blocks' in the jet stream. [online] Available at: https://www.science.org/content/article/why-does-weather-stall-new-theories-explain-enigmatic-blocks-jet-stream. Accessed October 2021.

(11)   Coumou et al (2018) The influence of Arctic amplification on mid-latitude summer circulation. *Nature Communications*. (9), pp:2959.

Figure 1-2 from Real World Globes – Investigating Jet Streams and Planetary (Rossby) Waves – by Will Robertson.

Figure 1-4 derived from R. Stull, 2017: Practical Meteorology

Chapter 2

(1) Macdonald, S.A. (1963) *From the Ground Up*. Aviation Publishers. Ottawa, Canada.

(2) Storelvmo, T., Tan, I. (2015) The Wegener-Bergeron-Findeisen process – Its discovery and vital importance for weather and climate. *Meteorologische Zeitrschrift*. [online] Available at: http://citeseerx.ist.psu.edu/viewdoc/download?doi=10.1.1.1061.7882&rep=rep1&type=pdf. Accessed: November 2021.

(3) Ludlum, F.H. (1966) Cumulus and Cumulonimbus Convection. *Tellus* XVIII, 4. pp: 687-98.

(4) Ludlum, F.H. (1963). Severe Storms: A Review. In: *Severe Local Storms*. Eds: Atlas, D. et al. American Meteorological Society, USA.

(5) WeatherWorks website (2021) Relative Humidity vs Dewpoint. [online] Available at: https://weatherworksinc.com/news/humidity-vs-dewpoint. Accessed: November 2021.

(6) Cotton, W., Bryan, G., van den Heever, S. (2011) Cumulonimbus Clouds and Severe Convective Storms. In: *Storm and Cloud Dynamics, The Dynamics of Clouds and Precipitating Mesoscale Systems*. Edited by William Cotton, George Bryan, Susan van den Heever, Volume 99, Pages 315-54.

(7) Allen, JT, et al (2019) Understanding Hail in the Earth System. *Reviews of Geophysics*. 10.1029/2019RG000665.

(8) Bilham, E.G., Relf, E.F. (1937) The Dynamics of Large Hailstones. *Quarterly Journal of Research Meteorology*. (63). Pp:149-162.

(9) Renick, J. H.,Maxwell J. B. (1977) Forecasting hailfall in Alberta. Hail: A Review of Hail Science and Hail Suppression, *Meteor. Monogr.*, No. 38,pp.145–151.

(10) Foote, G. B. (1984). A study of hail growth utilizing observed storm conditions. *Journal of Climate and Applied Meteorology*, 23(1), 84–101.

(11) SaskAdapt website (2021) Hail. [online] Available at: https://www.parc.ca/saskadapt/extreme-events/hail.html. Accessed: October 2021.

(12) Brimelow et al (2002) Modeling Maximum Hail Size in Alberta Thunderstorms. *Weather and Forecasting*. 17 (5) pp.1048–1062.

(13) NOAA website (2021) Storm Events Database. [online] Available at: https://www.ncdc.noaa.gov/stormevents.com. Accessed: Aug 29, 2021.

(14) Adams-Selin, R., Ziegler,C. (2016)Forecasting Hail Using a One-Dimensional Hail Growth Model with WRF. *American Meteorological Society*. Vol 144 (12).

(15) Brimelow et al (2017) The Changing Hail Threat over North America in response to Anthropogenic Climate Change. *Nature Climate Change*. Vol 7.

Figure 2-1 – from William Emery, Adriano Camps, in Introduction to Satellite Remote Sensing, 2017

Chapter 3

(1) Vonnegut, K. (1963) *The Cat's Cradle*. Holt Rinehart and Winston, USA.

(2) Petroleum service Company website (2017) Kurt Vonnegut, Cloud Seeding, and the Power of the Weather. [online] Available at: https://petroleumservicecompany.com/blog/ kurt-vonnegut-cloud-seeding-weather-control/. Accessed November 2021.

(3) Battan, L (1965) A view of cloud physics and weather modification in the Soviet Union. *Bulletin American Meteorological Society*. Vol. 46, No. 6.,pp:309-316.

(4)   Knowles, S. (2020) The Impact of Cloud Seeding On
        Small Grain Crops: Evidence from the North Dakota
        Cloud Modification Program. Unpublished. Michigan State
        University.

(5)   Kraus, T., Rennick, J. (1997) The New Alberta Hail
        Suppression Project. *Journal of Weather Modification*. vol 29
        (1). pp:100-105.

(6)   Gilbert et al (2019) Twenty Seasons of Airborne Hail
        Suppression In Alberta, Canada. Technical Report.
        [online] Available at: https://www.researchgate.net/
        publication/332684543_Twenty_Seasons_of_Airborne_
        Hail_Suppression_In_Alberta_Canada. Accessed
        November 2021.

(7)   Rivera, et al (2020) Sixty Years of Hail Suppression
        Activities in Mendoza, Argentina: Uncertainties, Gaps
        in Knowledge and Future Perspectives. Frontiers in
        Environmental Science. 8(45) pp: 1-6.

Chapter 4

(1)   Roberts, N., Woodbridge, J., Bevan, A., Palmisano, A.,
        Shennan, S., Asouti, E. (2018) Human responses and
        non-responses to climatic variations during the last
        Glacial-Interglacial transition in the eastern Mediterranean.
        Quaternary Reviews. Vol. 184, pp:47-67.

(2)   Erseck, K. (2012) Monocots Vs Dicots: What You Need
        To Know. [online] Available at: https://www.holganix.
        com/blog/monocots-vs-dicots-what-you-need-to-know.
        Accessed October 2021.

(3)   Bucholtz (2021) *Field to Flask – Fundamentals of Small Batch
        Distilling*. Wood dragon Books, Canada.

Chapter 5

(1) Pediaa website (2018) Difference Between Epigeal and Hypogeal Germination. [online] Availoable at: https://pediaa.com/difference-between-epigeal-and-hypogeal-germination. Accessed: November 2021.

(2) Australian Oil seeds website (2021) Understanding how Environment and Genotype Determine Time to Flowering in Canola and Indian Mustard. [online] Available at: http://www.australianoilseeds.com/__data/assets/pdf_file/0009/4500/Environment__and__genotype_-time_to_flowering.pdf. Accessed: Aug 27, 2021.

(3) D'Amico-Damiao,V., Carvalho,R.F. (2018) Cryptochrome-Related Abiotic Stress Responses in Plants. Frontiers in Plant Sciences. [online]. Available at: https://www.frontiersin.org/articles/10.3389/fpls.2018.01897/full. Accessed; Aug 27, 2021.

(4) SaskFlax website (2021) Early Flax History. [online] Available at: https://www.saskflax.com. Accessed: August 25, 2021.

(5) Rural Advantage website (2012) Management Tips for Flax Production. [online] Available at: http://ruraladvantage.org/wp-content/uploads/2012/04/management-tips-for-flax.pdf Accessed: September 28, 2021.

(6) Cullis, C. (2007) Flax. In: *Oilseeds*. Ed. C. Kole. Springer: New York.

(7) Flax Council of Canada website (2021) Growth and Development. [online] Available at: https://flaxcouncil.ca/growing-flax/chapters/growth-and-development. Accessed: Aug 26, 2021.

(8) Edirisinghe,P. (2016) Characterization of Flax Germplasm for Resistance to Fusarium Wilt Caused by Fusarium oxysporum f. sp. Lini. [online] Available at: https://harvest.

usask.ca/bitstream/handle/10388/7844/EDIRISINGHE-THESIS-2017.pdf?sequence=1&isAllowed=y. Accessed: October 2021.

(9) SaskCanola website (2021) The Canola Story. [online] Available at: https://www.saskcanola.com. Accessed: August 25, 2021.

(10) Snowdon,R, et al. (2007) Oilseed Rape. In: *Oilseeds*. Ed. C. Kole. Springer: New York.

(11) Canterra Seeds website (2014) The Difference Between Hybrid and Open Pollinated Canola. [online] Available at: https://canterra.com/blog/canola-seed/the-right-variety-for-the-job/ Accessed: September 26, 2021.

(12) Canola Council website (2021) Canola Growth Stages. [online] Available at: https://www.canolacouncil.org/canola-encyclopedia/growth-stages/#summary-of-canola-growth-stages. Accessed: Aug 27, 2021.

(13) Sask Mustard website (2021) Mustard Production Manual. [online] Available at: https://saskmustard.com/production-manual/plant-description/growth-stages/index.html. Accessed: Aug 29, 2021.

(14) Edwards, D. (2007) Indian Mustard. In: *Oilseeds*. Ed. C. Kole. Springer: New York.

(15) Liber, M., Duarter, I., Maia, A., Oliviera, H., (2021) The History of Lentil (Lens culinaris subsp. culinaris) Domestication and Spread as Revealed by Genotyping-by-Sequencing of Wild and Landrace Accessions. *Frontiers in Plant Science*. Vol 12, pp: 1-18.

(16) Sask Pulse website (2021) Lentil Staging Guide. [online] Available at: https://saskpulse.com/files/technical_documents/16SPG7820_Staging_Guides-Lentils_Web.pdf. Accessed: Aug 27, 2021.

(17) Croptrust website (2021) Lentils: History Through a Lens.

[online] Accessed: Aug 25, 2021. Available at: https://www.croptrust.org/impact-story/lentils-history-through-a-lens/

(18) Erskine et al (1990) Stages of Development in Lentil. *Experimental Agriculture*. 26(3), pp 297-306.

(19) Stefaniak, T., McPhee, K. (2015) Lentil. In: *Grain Legumes*. Ed. A De Ron. Springer: London.

(20) Hirst, K., (2019) Pea (Pisum sativum L.) Domestication - The History of Peas and Humans. [online] Available at: https://www.thoughtco.com/domestication-history-of-peas-169376. Accessed: Aug 25, 2021

(21) Warkentin, T. et al (2015) Pea. In: *Grain Legumes*. Ed. A De Ron. Springer: London.

(22) MacMillan, K. (2021). Soybean and Pulse Agronomy Program Lab 2019 and 2020 Annual Report.

(23) Millan, T. et al (2015) ChickPea. In: *Grain Legumes*. Ed. A De Ron. Springer: London.

(24) USA Pulses website (2021) Chickpeas. [online] Available at: https://www.usapulses.org/technical-manual/chapter-2-general-properties/chickpeas. Accessed: Aug 27, 2021.

(25) Sofi et al (2020) Chickpea. In: *Pulses*. Ed: A. Manickavasagan, Praveena Thirunathan. Springer: Switzerland.

(26) Kastner, J. (2018) *Corn: A History*. American Heritage Publishing: Maryland, USA.

(27) Randolph, L.F. (1976) Contributions of Wild Relatives of Maize to the Evolutionary History of Domesticated Maize: A Synthesis of Divergent Hypotheses I. *Economic Botany*. Vol. 30, No. 4. Pp: 321-45.

(28) Cai, H. (2006) Maize. In: *Cereals and Millets*. Ed: C. Kole. Springer: Berlin

(29) Bayer Canada Crop Science website (2021) Determining Corn Growth Stages. [online] Available at: https://

www.cropscience.bayer.ca/en/stories/2021/grow-your-knowledge/determining-corn-growth-stages. Accessed: Aug 27, 2021.

(30) Bayer Canada Crop Science website (2021) Cereal Staging. [online] Available at: https://www.cropscience.bayer.ca/~/media/Bayer%20CropScience/Country-Canada-Internet/Growers%20Tools/Resources-and-Guides/cereal-staging-guide.ashx Accessed: Aug 27, 2021.

(31) Iowa State Extension and Outreach website (2021) Corn Growth Stages. [online] Available at: https://crops.extension.iastate.edu/encyclopedia/corn-growth-stages. Accessed: October 2021.

(32) New World Encyclopedia (2021) [online] Available at: https://www.newworldencyclopedia.org/entry/Cereal. Accessed: October 2021.

(33) Einkorn website (2021) The History of Einkorn. [online] Available at: https://www.einkorn.com/einkorn-history/. Accessed: October 2021.

(34) Australian Grain Growers website (2021) The History of Oats. [online] Available at: https://www.graingrowers.com.au/the-history-of-oats/. Accessed: October 2021.

(35) Zadoks, J.C., Chang, T.T. & Konzak, C.F. (1974) A decimal code for the growth stages of cereals. *Weed Research* 14, 415-2.

(36) Herbek,J., Lee, C. (2021)Growth and Development. In: *A Comprehensive Guide to Wheat Management in Kentucky.* [online] Available at: http://www2.ca.uky.edu/agcomm/pubs/id/id125/02.pdf. Accessed: October 2021.

(37) Konduru, S. et al (2008) The Global Economic Impacts of Roundup Ready Soybeans. In: *Genetics and Genomics of Soybean.* Ed: G. Stacey. Springer: New York.

(38) Lee et al (2007) Soybean. In: *Oilseeds.* Ed. C. Kole. Springer:Berlin.

(39) deKalb website (2021) Soybean Growth and Development. [online] Available at: https://www.dekalbasgrowdeltapine.com/en-us/agronomy/soybean-growth-stages.html. Accessed: Aug 28, 2021.

(40) Iowa State University (2021) Soybean Growth Stages. [online] Available at: https://crops.extension.iastate.edu/soybean/production_growthstages.html. Accessed: October 2021.

(41) University of Minnesota (2021) Soybean Growth Stages. [online] Available at: https://extension.umn.edu/growing-soybean/soybean-growth-stages. Accessed: October 2021.

(42) Grains Canada website (2021) Canary Seed: A Novel Cereal from the Canadian Prairies. [online]. Available at: https://grainscanada.gc.ca/en/grain-research/scientific-reports/canary/pdf/canary-en.pdf. Accessed: September 28, 2021.

(43) Canary seed website (2021) About Canary Seed. [online]. Available at: https://www.canaryseed.ca/about.html. Accessed: Aug 29, 2021.

(44) Abdel-Aal, ESM, et al (2011) Microstructure and nutrient composition of hairless canary seed and its potential as a blending flour for food use. *Food Chemistry* 125, pp 410–416.

(45) Abdel-Aal, ESM, (2021) Nutritional and functional attributes of hairless canary seed groats and components and their potential as functional ingredients. *Trends in Food Science & Technology* 111,pp: 680–687.

(46) Alberta Pulse website (2021) Faba Beans. [online] Available at: https://albertapulse.com/growing-faba-beans/ Accessed: September 28, 2021.

(47) GRDC website (2021) https://grdc.com.au/__data/assets/pdf_file/0025/369142/GrowNote-Faba-Bean-West-4-Physiology.pdf.

(48) Crepon, K., et al (2010) Nutritional value of faba bean

(Vicia faba L.) seeds for feed and food. *Field Crops Research* 115 (2010) 329–339

(49) Juszczak,D., Wesołowski, M. (2011) Phenological Phases of Buckwheat (Fagopyrum esculentum Mnch.)In The Primary And Secondary Crop Depending On Seeding Rate. *Acta Agrobotanica* , Vol. 64 (4): 213–226.

(50) NDSU website (2019) Buckwheat Production. [online] Available at: https://www.ag.ndsu.edu/publications/crops/ buckwheat-production. Accessed: October, 2021.

(51) Arendt,E., Zannini, E. (2013) Buckwheat. In: *Cereal grains for the food and beverage industries*. Woodhead Publishing, New York.

(52) USDA website (2010) *Small Grains Loss Adjustment Standards Handbook*. [online] Available at: https://legacy.rma.usda. gov/handbooks/25000/2010/10_25430-1h.pdf. Accessed: Oct 2021.

(53) Schneiter,A., Miller, J.F., (1981) Description of Sunflower Growth Stages. *Crop Sci*.11: 635-638.

(54) Grains SA website (2019) Sunflower and its stages of development. [online] Available at: https://www.grainsa. co.za/sunflowers-and-its-stages-of-development. Accessed: October 2021.

(55) Iowa State University Extension website (2021) Sunflower. [online] Available at: https://www.extension.iastate.edu/ alternativeag/cropproduction/sunflower.html. Accessed October 2021.

(56) Yara website (2021) Growth and Development of the sugar beet. [online] Available at: https://www.yara.co.uk/crop-nutrition/sugar-beet/growth-and-development-of-sugar-beet/. Accessed: October 2021.

(57) Montana State University extension website (2021) Sugar Beet. [online] Available at: https://pspp.msuextension.org/ Sugarbeets.html. Accessed: October 2021.

(58) Cattanach, A.W. et al (1991) Sugar Beets. [online] Available

at: https://hort.purdue.edu/newcrop/afcm/sugarbeet. html. Accessed: October 2021.

Chapter 6

(1)   Freeman, B.C., Beattie, G.A.(2008). An Overview of Plant Defenses against Pathogens and Herbivores. The Plant Health Instructor.

Figure 6-4 from https://biologydictionary.net.

Chapter 7

(1)   Klages, K.H. (1933) The effects of simulated hail injuries on flax. *Agronomy Journal.*

(2)   Soine, O.C. (1970). Effect of simulated hail damage on flax. Agricultural Experiment Station, University of Minnesota, Miscellaneous Report 92.

(3)   Casa et al (1999) Environmental effects on linseed (Linum usitatissimum L.) yield and growth of flax at different stand densities. *European Journal of Agronomy* 11, pp: 267–278.

(4)   McGregor, D.I. (1987) Effect of Plant Density on Development and Yield of Rapeseed and its Significance to Recovery from Hail Injury. *Can. J. Plant Sci.* 67: 43-51.

(5)   Syrovy, L.D. et al (2016) Yield Response to Early Defoliation in Spring-Planted Canola. *Crop Science*, vol. 56, pp:1981-1987

(6)   Vollmer, J. (2019) Simulated Hail Damage on Spring Canola (BRASSICA NAPUS L.): Nonuniform Stand Reduction and Cut-Off. Un-published. [online] Available at: https://library.ndsu.edu/ir/bitstream/ handle/10365/29780/Vollmer_ndsu_0157N_12325. pdf?isAllowed=y&sequence=1. Accessed: October 2021.

(7)   Sask Pulse website (2021) Environmental Stresses in Pulses.

[online] Available at: https://saskpulse.com/files/technical_documents/190603_Environmental_Damage_to_Pulses-compressed.pdf. Accessed: Aug 29, 2021.

(8)   Bueckert, R.A. (2011) Simulated hail damage and yield reduction in lentil. *Can. J. Plant Sci.* 91: pp: 117-124.

(9)   Miller, D. G. and Muehlbauer, F. J. (1984). Stem excision as a means of simulating hail injury on 'Alaska' peas. *Agron. J.* 76: pp:1003-1005.

(10)   Li,L.et al (2010) Shading, Defoliation and Light Enrichment Effects on Chickpea in Northern Latitudes. *J. Agronomy & Crop Science.*

(11)   Battaglia et al (2019) Hail Damage Impacts on Corn Productivity: A Review. *Crop Science*, vol. 59, pp:1-14.

(12)   Eldredge, John C. (1935) "The effect of injury in imitation of hail damage on the development of the corn plant," *Research Bulletin* (Iowa Agriculture and Home Economics Experiment Station): Vol. 16 : No. 185 , Article 1.

(13)   Roth. G.W., Lauer, J.G. (2008) Impact of Defoliation on Corn Forage Quality. *Agronomy Journal*, 100 (3), pp: 651-7.

(14)   Eldredge, John C. (1937) "The effect of injury in imitation of hail damage on the development of small grain," *Research Bulletin* (Iowa Agriculture and Home Economics Experiment Station): Vol. 19 : No. 219 , Article 1.

(15)   Stone, E.G. (1955 The Effect of Simulated Hail Damage During Various Stages of Growth on the Yield of Winter Wheat in Oklahoma. Unpublished. [online] Available at: https://hdl.handle.net/11244/32699

(16)   Sanchez, et al (1996) Crop Damage: The Hail Size Factor. *Journal of Applied Meteorology.* Vol. 35, No. 9, pp. 1535-1541.

(17)   Bayer Crop Science website (2021) Wheat rust diseases. [online] Available at: https://www.cropscience.bayer.us/learning-center/articles/wheat-rust-diseases#phcontent_6_

divAccordion. Accessed: October 2021.

(18) Klein, R, Shapiro, C (2011) Evaluating Hail Damage to Soybeans. [online] Available at: https://extensionpublications.unl.edu/assets/pdf/ec128.pdf. Accessed: October 2021.

(19) Iowa State University Extension website (2016) Hail on Soybean in Iowa.

(20) USDA Procedures (2010) Small Grains Loss Adjustment Standards Handbook. [online] Available at: https://legacy.rma.usda.gov/handbooks/25000/2010/10_25430-1.pdf. Accessed: November 2021.

(21) National Sunflower Assn website (2021) [online] Available at: https://www.sunflowernsa.com/uploads/25/hail-damage-adjustment-guide.pdf. Accessed: October 2021.

(22) Jones et al (1955) The Effects of Defoliation and Loss of stand Upon Yield of Sugar Beet. *Annals of Applied Biology*, 43: 63-70.

(23) Muro et al (1998) Defoliation Timing and Severity in Sugar Beet. *Agron. J.* 90, pp:800-804

Chapter 8

(1) Noreen,S. et al (2018) Foliar Application of Micronutrients in Mitigating Abiotic Stress in Crop Plants. In: *Plant Nutrients and Abiotic Stress Tolerance*. Ed: Hasanuzzaman et al, Springer: Singapore.

(2) Aziz, et al (2019) Foliar application of micronutrients enhances crop stand, yield and the biofortification essential for human health of different wheat cultivars. *Journal of Integrative Agriculture*, 18(6), pp: 1369–1378.

(3) Stewart et al (2020) Foliar Micronutrient Application for High-Yield Maize. Foliar Micronutrient Application for High-Yield Maize. *Agronomy* (10), 1946, pp:1-18.

(4) Wang, et al (2015) Foliar Zinc, Nitrogen, and Phosphorus Application Effects on Micronutrient Concentrations in Winter Wheat. *Agronomy Journal*. 107 (1) pp: 61–70.

(5) Gaj, et al (2020) The Effect of Potassium and Micronutrient Foliar Fertilisation on the Content and Accumulation of Microelements, Yield and Quality Parameters of Potato Tubers. *Agriculture*, 10 (530). pp:1-14.

(6) Venugopalan et al (2021) The Response of Lentil (Lens culinaris Medik.) to Soil Moisture and Heat Stress Under Different Dates of Sowing and Foliar Application of Micronutrients. *Frontiers in Plant Science*. Vol 12. pp:1-15.

(7) Hadi et al (2020) Effect of Foliar Application of Some Micronutrients at Two Growth Stages on Growth, Yield and Yield Components of Two Bread Wheat (Triticum aestivum L.) Varieties. *ZANCO Journal of Pure and Applied Sciences*. 32 (5): 186-195

(8) Knell M (2010) Nanotechnology and the sixth technological revolution.In: Cozzens SE, Wetmore JM (eds) *Nanotechnology and the challenges of equity, equality and development*. Springer, Dordrecht, pp:127–143

Chapter 9

(1) Fisher, M. et al (2012) Emerging fungal threats to animal, plant and ecosystem health. *Nature*, 484 (7393)

(2) Holbein, J. et al (2016) Plant basal resistance to nematodes: an update. *Journal of Experimental Botany*, Volume 67, Issue 7, pp: 2049–2061.

(3) Freeman, B.C., Beattie, G.A. (2008). An Overview of Plant Defenses against Pathogens and Herbivores. *The Plant Health Instructor*. Iowa State University.

(4) Prasad, E. (2021) Fungicide Market Expected to Reach $19.5 billion by 2027. [online] Available at: https://www.

alliedmarketresearch.com/press-release/global-fungicides-market-is-expected-to-reach-16-2-billion-by-2020-allied-market-research.html. Accessed November 2021.

Figure 9-1 credit to Shutterstock image number 1105352483

Chapter 12

(1)   FAA website (2021) GPS How it Works. [online] Available at: https://www.faa.gov/about/office_org/headquarters_offices/ato/service_units/techops/navservices/gnss/gps/howitworks. Accessed December 2021.

(2)   GIS Geography (2021) WGS 84 System. [online] Available at: https://gisgeography.com/wgs84-world-geodetic-system. Accessed: December 2021.

(3)   ThoughtCo website (2021) Visible Light Spectrum. [online] Available at: https://www.thoughtco.com/the-visible-light-spectrum-2699036. Accessed December 2021.

(4)   ExplainThatStuff website (2021) 35mm Film Cameras. [online] Available at: https://www.explainthatstuff.com/how-film-cameras-work.html. Accessed: December 2021.

(5)   How Stuff Works website (2021) Digital Cameras. [online] Available at: https://electronics.howstuffworks.com/cameras-photography/digital/digital-camera4.html. Accessed: December 2021.

(6)   Jin, X. et al (2017) Estimates of plant density of wheat crops at emergence from very low altitude UAV imagery. *Remote Sensing of Environment.* Vol 198, pp: 105-114.

(7)   Yao, X. et al (2017) Estimation of Wheat LAI at Middle to High Levels Using Unmanned Aerial Vehicle Narrowband Multispectral Imagery. *Remote Sens.* 9(1304), pp 1-14.

(8)   Zhou, J. et al (2016) Aerial multispectral imaging for crop hail damage assessment in potato. *Computers and Electronics in*

Agriculture. 127, pp: 406–412.

Figure 12-1 credit to Fouda, Y. (2014) Urban planning advanced analytic techniques and the human role: A challenge or a complement. *Alexandria Engineering Journal*, Vol. 43, No. 6.

www.ingramcontent.com/pod-product-compliance
Lightning Source LLC
Chambersburg PA
CBHW072300210326
41519CB00057B/2354